Tributes
Volume 28

Conceptual Clarifications
Tributes to Patrick Suppes (1922-2014)

Volume 18
Insolubles and Consequences. Essays in Honour of Stephen Read.
Catarina Dutilh Novaes and Ole Thomassen Hjortland, eds.

Volume 19
From Quantification to Conversation. Festschrift for Robin Cooper on the occasion of his 65th birthday
Staffan Larsson and Lars Borin, eds.

Volume 20
The Goals of Cognition. Essays in Honour of Cristiano Castelfranchi
Fabio Paglieri, Luca Tummolini, Rino Falcone and Maria Miceli, eds.

Volume 21
From Knowledge Representation to Argumentation in AI, Law and Policy Making. A Festschrift in Honour of Trevor Bench-Capon on the Occasion of his 60th Birthday
Katie Atkinson, Henry Prakken and Adam Wyner, eds.

Volume 22
Foundational Adventures. Essays in Honour of Harvey M. Friedman
Neil Tennant, ed.

Volume 23
Infinity, Computability, and Metamathematics. Festschrift celebrating the 60th birthdays of Peter Koepke and Philip Welch
Stefan Geschke, Benedikt Löwe and Philipp Schlicht, eds.

Volume 24
Modestly Radical or Radically Modest. Festschrift for Jean Paul Van Bendegem on the Occasion of his 60th Birthday
Patrick Allo and Bart Van Kerkhove, eds.

Volume 25
The Facts Matter. Essays on Logic and Cognition in Honour of Rineke Verbrugge
Sujata Ghosh and Jakub Szymanik, eds.

Volume 26
Learning and Inferring. Festschrift for Alejandro C. Frery on the Occasion of his 55th Birrthday
Bruno Lopes and Talita Perciano, eds.

Volume 27
Why is this a Proof? Festschrift for Luiz Carlos Pereira
Edward Hermann Haeusler, Wagner de Campos Sanz and Bruno Lopes, eds.

Volume 28
Conceptual Clarifications. Tributes to Patrick Suppes (1922-2014)
Jean-Yves Béziau, Décio Krause and Jonas R. Becker Arenhart, eds.

Tributes Series Editor
Dov Gabbay   dov.gabbay@kcl.ac.uk

Conceptual Clarifications
Tributes to Patrick Suppes (1922-2014)

edited by
Jean-Yves Béziau,
Décio Krause
and
Jonas R. Becker Arenhart

© Individual authors and College Publications 2015. All rights reserved.

ISBN 978-1-84890-188-9

College Publications
Scientific Director: Dov Gabbay
Managing Director: Jane Spurr

http://www.collegepublications.co.uk

Cover design by Laraine Welch

Printed by Lightning Source, Milton Keynes, UK

All rights reserved. No part of this publication may be reproduced, stored in a retrieval system or transmitted in any form, or by any means, electronic, mechanical, photocopying, recording or otherwise without prior permission, in writing, from the publisher.

CONTENTS

Notes on the Contributors   iii

JEAN-YVES BÉZIAU, DÉCIO KRAUSE AND JONAS ARENHART
Conceptual Clarifications, Tributes to Patrick Suppes (1922 - 2014)
A Preface   v

J. ACACIO DE BARROS, GARY OAS AND PATRICK SUPPES
Negative Probabilities and Counterfactual Reasoning on the Double-slit Experiment   1

DÉCIO KRAUSE AND JONAS R. BECKER ARENHART
Logical Reflections on the Semantic Approach   31

STEVEN FRENCH
Between Weasels and Hybrids: What Does
the Applicability of Mathematics Tell us about Ontology?   63

SILVIA HARING AND PAUL WEINGARTNER
Environment, Action Space and Quality of Life:
An Attempt for Conceptual Clarification   87

F.A. MULLER
Circumveiloped by Obscuritads:
The nature of interpretation in quantum mechanics, hermeneutic circles and physical reality, with cameos of James Joyce and Jacques Derrida   107

NEWTON C. A. DA COSTA AND OTÁVIO BUENO
Structures in Science and Metaphysics   137

GERGELY SZÉKELY
What Algebraic Properties of Quantities
Are Needed to Model Accelerated Observers
in Relativity Theory   161

ARNOLD KOSLOW
Laws, Accidental Generalities, and the Lotze Uniformity Condition 175

JEAN-YVES BEZIAU
Modeling Causality   187

DAVID MILLER
Reconditioning the Conditional   205

PAUL WEINGARTNER
A 6-Valued Calculus which Avoids the
Paradoxes of Deontic Logic   217

ANNE FAGOT-LARGEAULT
The Psychiatrist's Dilemmas   229

Notes on the Contributors

List of Contributors

Jonas R. Becker Arenhart: Department of Philosophy, Federal University of Santa Catarina.

J. Acacio de Barros: Liberal Studies Program, San Francisco State University.

Jean-Yves Béziau: Department of Philosophy, Federal University of Rio de Janeiro and Visiting Professor, University of California, San Diego.

Otávio Bueno: Department of Philosophy, Miami University.

Newton C. A. da Costa: Department of Philosophy, Federal University of Santa Catarina.

Anne Fagot-Largeault: School of Philosophy, Religion and History of Science, College de France & Academie des Sciences.

Steven French: School of Philosophy, Religion and History of Science, University of Leeds.

Silvia Haring: Department of Psychology, University of Salzburg.

Arnold Koslow: The Graduate Center, Cuny.

Décio Krause: Department of Philosophy, Federal University of Santa Catarina.

David Miller: Department of Philosophy, University of Warwick.

F.A. Muller: Faculty of Philosophy, Erasmus University Rotterdam, and Institute for the History and Foundations of Science, Department of Physics & Astronomy, Utrecht University.

Gary Oas: EGPY – Education Program for Gifted Youth, Stanford University, Stanford.

Patrick Suppes: CSLI – Centre for the Study of Language and Information, EGPY – Education Program for Gifted Youth, Stanford University.

Gergely Székely: Alfréd Rényi Institute of Mathematics, Hungarian Academy of Sciences.

Paul Weingartner: Department of Philosophy, University of Salzburg.

Conceptual Clarifications, Tributes to Patrick Suppes (1922 - 2014)
A Preface

JEAN-YVES BÉZIAU, DÉCIO KRAUSE AND JONAS ARENHART

This is a collection of papers dedicated to the memory of Patrick Colonel Suppes (1922-2014) by people who have been closely connected with him and his work. It was first thought of as a kind of natural follow up of a special issue of *Synthese* (Volume 154, Issue 3, February 2007) edited by the first two editors of this book commemorating the 80th birthday of Pat Suppes. The title of this issue was *New Trends in the Foundations of Science*. The subtitle of the present volume reflects — in some measure — the situation in which it must be presented to the public, given Suppes' passing away during its production.

So, with this laudatory intention in mind, the whole volume was first thought of as a Tribute to Suppes due to the occasion of his 90th anniversary. We shall not speculate on the reasons why people think that anniversaries should be commemorated with a volume — when they are commemorated at all — from ten to ten years (why don't we find a homage to someone's 87th birthday?). The first plan, anyway, was that the volume should be a well deserved homage to Suppes, commemorating his long life of productive interaction, influence, and direct collaboration with a great variety of researchers. This did not mean that the homage took it as a fact that Suppes' work was finished, or that Suppes was only an influent philosopher of the past that had nothing else to say: he was still active and developing influential new ideas in many fields. In fact, one of the papers in this collection is co-authored by Suppes himself. So, the idea was clearly not merely to praise someone, but to continue an ongoing debate on some of the issues discussed in the work of Suppes himself.

The work of Suppes touches many different areas, ranging from meteorology to physics, through logic, mathematics, psychology, neuroscience, education, painting, but he was first of all and/or above all a philosopher, always questioning, but not in vain. Part of Suppes' research in the foundations of science culminates in his book *Representation and invariance of scientific structures* (CSLI, Stanford, 2002). This book is a synthesis showing clearly the relations between all the topics he has investigated. There are not many philosophers who can be proud of having written influential math textbooks, contributed decisively to the philosophical foundations of science, helped to develop research in real labs, and much more. Suppes is such a singular figure. Not only did he think about some of those important issues, but also helped to bring some of them about, as in the particular case of computer-assisted learning; perhaps, that is the aspect of his work for which he will be reminded by most of the public out of philosophical circles (again, not many philosophers can be proud of that too).

Since the range of interest of Suppes was very broad, so is the variety of topics dealt with in this volume. In fact, the work of a researcher is certainly not limited to his own writings, but has to be appreciated also through the work of the people he has been working with and influenced. From this point of view the work of Suppes is very impressive, and the present book contributes to show that. This feature —wide range of interest and great competence to deal with such interests properly — appears very clearly in this collection; on the one hand the volume clearly bears the sign of Suppes' influence and wide ranging interests as a scholar: most authors have had direct contact and have felt the need to discuss Suppes ideas; the themes are varied and their main thread is..., well, their relation, direct or indirect, with the work of Suppes. On the other hand, with such a long life span and such a great variety of interests, the volume should certainly contain a broad spectrum of themes that may seem at first to lack a unifying thread. In fact, as readers may notice, it is not possible to contemplate — not even in a whole volume — the whole spectrum of areas of interest featuring in the works of Suppes. This volume illustrates this: hardly two papers deal with the same subject. Suppes was such an apt thinker as to see unity in many disparate areas.

Most of the authors of the papers collected here had, at first, the expectation that their contribution would be indeed a further step in their fruitful interaction with Suppes; most of the papers develop themes that were among Suppes' areas of direct research, in the present or in some moment of his life. Anyway, the celebratory tone of the volume was not meant to obscure the fact that another round of stimulating intelectual dialog with Suppes himself was expected. It is not everyday that we meet someone with such a great expertise in so many areas of intelectual investigation, so, the opportunity should not be missed by anyone.

However, as we have mentioned, during the preparation of the volume Suppes passed away. This happening made the present volume into a posthumous homage celebrating his work and influence. The fact that it is now a posthumous homage does not change that Suppes' contribution is a lively one; the debate should go on following the steps of Suppes' fruitful contributions. We are proud of having a paper by Suppes himself and his collaborators in our volume. Not only was he a philosopher deserving to be praised, but was also a philosopher in the scientific sense of the word. For those investigating the nature of science there is no stoping point, science always presents another challenge and the philosopher must be always on guard. Suppes did that in an exemplary fashion, contributing to both science and philosophy.

The editors would like the take the opportunity to thank all the authors that contributed to this volume. The quality of the papers here gathered certainly are a fair tribute to the greatness of the philosopher Suppes was. We would also like to thank the authors for their immense patience it took for the production of the volume. As we know, academic pressures and other contingencies sometimes provide for unwanted and unexpected delays. However, at last, it is ready and out for the public judgement. May the volume meet its public and help to enlighten the themes here touched on, themes that were dear to Suppes, and it will have reached its aim.

Federal University of Rio de Janeiro - UFRJ
Rio de Janeiro, RJ
BRAZIL

Federal University of Santa Catarina - UFSC
Florianópolis, SC
BRAZIL

Negative Probabilities and Counterfactual Reasoning on the Double-slit Experiment

J. ACACIO DE BARROS, GARY OAS AND PATRICK SUPPES

ABSTRACT. In this paper we attempt to establish a theory of negative (quasi) probability distributions from fundamental principles and apply it to the study of the double-slit experiment in quantum mechanics. We do so in a way that preserves the main conceptual issues intact but allow for a clearer analysis, by representing the double-slit experiment in terms of the Mach-Zehnder interferometer, and show that the main features of quantum systems relevant to the double-slit are present also in the Mach-Zehnder. This converts the problem from a continuous to a discrete random variable representation. We then show that, for the Mach-Zehnder interferometer, negative probabilities do not exist that are consistent with interference and which-path information, contrary to what Feynman believed. However, consistent with Scully et al.'s experiment, if we reduce the amount of experimental information about the system and rely on counterfactual reasoning, a joint negative probability distribution can be constructed for the Mach-Zehnder experiment.

1 Introduction

Ever since its inception, quantum mechanics has not ceased to perplex physicists with its counter-intuitive descriptions of nature. For instance, in their famous paper Albert Einstein, Boris Podolsky, and Nathan Rosen (EPR) argued the incompleteness of the quantum mechanical description [Einstein *et al.*, 1935]. At the core of their argument was the superposition of two wavefunctions where properties of two particles far apart, A and B, were highly correlated. Since both particles are spatially separated, EPR argued that a measurement on A should not affect B. Therefore, we could use the correlation and a measurement on A to infer the value of a property in B *without disturbing* it. Thus, concluded EPR, the quantum mechanical description of nature had to be incomplete, as it did not allow the values of a property to be fixed before an experiment was performed.

In 1964, John Bell showed that not only quantum mechanics was incomplete, but also that a complete description of physical reality such as the one espoused by EPR was incompatible with quantum mechanical predictions [Bell, 1964]. Later on, Alain Aspect, Jean Dalibard, and Gérard Roger, in an impressive and technically challenging experiment, obtained correlation measurements between measurement events separated by a spacelike interval. Their correlations supported quantum mechanics, in disagreement with EPR's metaphysical views [Aspect et al., 1981; Aspect et al., 1982]. Other puzzling results followed, like the Kochen-Specker theorem [Kochen and Specker, 1967; Kochen and Specker, 1975], Wheeler's delayed choice experiment [Wheeler, 1978; Jacques et al., 2007], the quantum eraser [Scully and Druhl, 1982], and the Greenberger-Horne-Zeilinger paradox [Greenberger et al., 1989; de Barros and Suppes, 2000; de Barros and Suppes, 2001], to mention a few.

What most of the above examples have in common (with the exception of Kochen-Specker) is that they all use superpositions of quantum states. There is nothing more puzzling in quantum mechanics than the fact that a given system can be in a state with two incompatible properties simultaneously "present." For example, let $\hat{O}$ be an observable corresponding to a property **O**, and $|o_1\rangle$ and $|o_2\rangle$ two eigenstates of $\hat{O}$ with eigenvalues o_1 and o_2. If the quantum system is in the state $|o_1\rangle$, we may say that it has an objective property **O** and its value is o_1, in the sense that if we measure this system for property **O**, the outcome of this measurement will be o_1 with probability one. Similarly if the system is in the state $|o_2\rangle$. However, it is also in principle possible to prepare a system in a quantum state that is the superposition of $|o_1\rangle$ and $|o_2\rangle$, i.e. $c_1|o_1\rangle + c_2|o_2\rangle$, where c_1 and c_2 are any complex numbers satisfying the constraint $|c_1|^2 + |c_2|^2 = 1$. At a first glance, this may not seem puzzling, as it is just telling us that the system is perhaps in a state where the value of **O** is unknown, except that it can either be o_1 (o_2) with probability $|c_1|^2$ ($|c_2|^2$). The perplexing aspects of superpositions come from the study of properties such as **O** taken in conjunction with other properties, say **O'**, in cases where their corresponding observables, $\hat{O}$ and $\hat{O}'$, do not commute ($[\hat{O}, \hat{O}'] \neq 0$).

There is perhaps no simpler context in which the superposition mystery reveals itself than single photon interference, realized in the double-slit experiment. In fact, in his Lectures in Physics, Richard Feynman famously claimed that this experiment contains the *only* mystery of quantum mechanics [Feynman et al., 2011]. Although there are other mysteries, as [Silverman, 1995] pointed out, the double-slit experiment provides us with an understanding of some key aspects of quantum mechanics. It may be the true mystery of quantum mechanics lies in the idea that a property is not the same in different contexts [Dzhafarov and Kujala, 2013a; Markiewicz et al., 2013; Howard et al., 2014; de Barros et al., 2014], and that such contexts (perhaps freely) chosen by an observer can be spacelike separated [Bell, 1964; Bell, 1966; Greenberger et al., 1989; de Barros and Suppes, 2000; de Barros and Suppes, 2001; Dzhafarov and Kujala, 2014d].

One of the main "disturbing" mysteries of the double-slit system, as described by Feynman, is the non-monotonic character of the probabilities of detection. This was exactly what motivated Feynman to use negative probabilities to describe quantum systems [Feynman, 1987]. However, as Feynman remarked, such an approach did not seem to provide any new insights into quantum mechanics.

It is our goal here to show that we can indeed gain some insight by using negative probabilities. This paper is organized the following way. First, we introduce in Section 2 a simplified version of the double-slit experiment in the form of the Mach-Zehnder interferometer. Then in Section 3, we present a theory of negative probabilities. We then show, in Section 4, that proper probability measures do not exist for the simultaneous measurements of the particle- and wave-like properties, but negative probabilities do under certain counterfactual conditions that are often assumed in experimental analyses.

2 The Mystery of the double-slit Experiment

In this section we reproduce Feynman's discussion of the double-slit experiment, and why he considered it mysterious [Feynman et al., 2011]. Feynman's argument involves the idea that classically we think of systems in terms of two distinct and incompatible concepts, particles or waves[1]. Such concepts are incompatible because particles are localized and waves are not. To see this, let us start with a point particle. In classical mechanics, the main characteristic of particles is that they are objects localized in space, and therefore can only interact with other systems that are present in their localized position. For example, let us consider a particle P whose position at time t is described by the position vector $\mathbf{r}_P(t)$. At time t_0, P can only interact with another physical entity that is also at $\mathbf{r}_P(t_0)$, either another particle S such that $\mathbf{r}_S(t_0) = \mathbf{r}_P(t_0)$ or a field that is nonzero at $\mathbf{r}_P(t_0)$. For instance, when a particle is subject to no external fields (of course an idealization), such as gravity, it travels in a straight line at constant speed, since no interac-

---

[1]You could also have fields, but in the context of the double-slit experiment, as it will become clear later, the relevant property of a field would be its spatial oscillations as a wave.

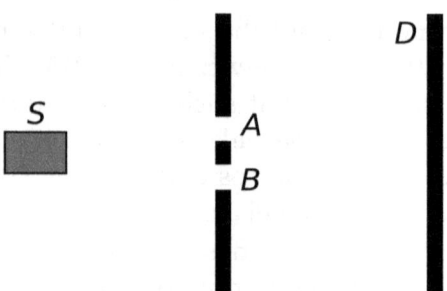

Figure 1. Schematics of the double-slit experiment. A source S emits a physical object towards a barrier, where two slits, A and B, are cut to allow for its passage. Then, at a screen D, the object is detected.

tion is present. If this particle then collides with another particle, say a constituent of a wall placed in the way of the original particle, an interaction will appear[2]. But, as soon as the particle looses contact with the wall, the interaction ceases to exist. In other words, particles interact locally.

The second basic concept is that of waves. Historically, the physics describing a point particle was extended to include the description of continuous media, and, more important to our current discussion, the vibrations of such media in the form of waves. Waves, therefore, were considered vibrations of a medium made out of several point particles, and the local interactions between two neighboring particles would allow for a perturbation in one point of the medium to be propagated to another point of the medium. Without going into the discussion of the particulars of electromagnetic waves, the main point is that because a single particle has an infinite number of neighbors, a disturbance on its position propagates to *all* of its neighbors and to *all* directions. Thus, the effect of such perturbation on particle S' belonging to this medium due to a perturbation on particle P' does not depend on the direct contact of S' and P'. More importantly, such effect depends not only on P', but also possibly on all other particles that make up the medium, and also on all interactions or boundary conditions that such particles need to satisfy. In other words, waves interact non-locally.

Going back to the double-slit experiment, let us analyze it from the point of view of what one should expect to happen were it being modeled with particles or with waves. Figure 1 shows a typical double-slit setup. We start with particles. Let us assume that S sends particles

---

[2]There is, of course, the obvious issue of how could this interaction be relevant, given that it would occur with probability zero.

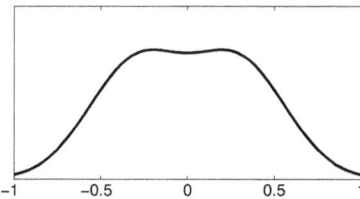

Figure 2. Probability density of observation for the double-slit experiment, assuming a particle model.

in random directions. A particle leaving S would interact only locally with the barrier and nothing else. This means that in between S and the barrier, this particle travels along a straight line. Once it reaches the barrier, it either goes through one of the slits and reaches D, or is reflected back (or absorbed, depending on the barrier's properties). While going through a slit, the particle may interact with the walls, perhaps bouncing off of it, and therefore causing some scattering from the direct path between S and D. Thus, if we run this experiment many times, we should expect the observed probability distribution of particles on D to be somewhat like as depicted in Figure 2. The resulting probability is simply the (normalized) sum of the probability of a particle going through slit A and B.

A wave analysis of the experiment shows something quite different. First, a wave is the result of a perturbation of a medium. In this case, the source S disturbs the medium, and such disturbance is propagated in all directions. One characteristic of such propagation is that its speed is dependent on the medium, and for the double slit experiment this is reflected in the arrival of a wave crest (or valley) in A at the same time that a crest (or valley) arrives in B. If A and B are small compared to the wavelengths, we can think of them as secondary wave sources oscillating in phase. Thus, when they arrive at D, in some places they will be in phase, whereas in other places they will be out of phase. The result is the constructive and destructive interference pattern that we know for waves, shown in Figure 3.

Now for the puzzling aspect of it all: quantum systems, e.g. electrons, have particle-like characteristics while at the same time being prone to wave-like interference. Here is how those aspects manifest themselves. Experimentally, electrons are particles. We know this because they are localized, in the sense that when we measure an electron, it shows up as a point in a fluorescent screen or a localized detector. This is to be contrasted with waves, which are spread out, and

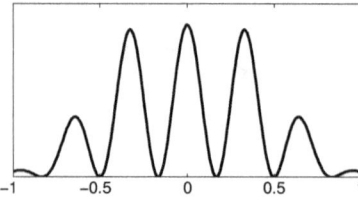

Figure 3. Intensity (arbitrary units) of the wave at D for the double-slit experiment.

therefore have measurable components (i.e., momentum and energy) in more than one place. Thus, as particles, when an electron leaves the source S, it will either go to slit A or B. Since its interactions are solely local, if it goes through, say, A, it can only interact with A, and not with B. Therefore, to a particle, it makes no difference at all if slit B is open or not when it goes through A, and *vice versa*. This is why we should expect a distribution like that on Figure 2. The disturbing fact is that an electron, if both slits are open, satisfies, after many runs, the distribution given in Figure 3. So, here is the puzzle. For a wave, the intensity is zero at several points, e.g. at position 0.5. How can this zero intensity be understood? How can the particle "know" (in the sense of interacting) about B if only interacting locally with A? To make this point even clearer, we will examine this question in detail in Section 4, using a simplified case of the double-slit experiment, the Mach-Zehnder interferometer, together with a framework of extended probabilities. But first, let us examine the concept of negative probabilities.

3 Negative Probabilities

In this Section we lay out the main relevant results and definitions for negative probabilities. We start by first defining it in a way that is related to Kolmogorov's [Kolmogorov, 1950] probability measure. We then prove some simple but relevant results.

Definition 1. Let Ω be a finite set, $\mathcal{F}$ an algebra over Ω, and p a real-valued functions, $p : \mathcal{F} \to \mathbf{R}$. Then $(\Omega, \mathcal{F}, p)$ is a probability space, and p a probability measure, if and only if:

K1. $0 \leq p(\{\omega_i\}) \leq 1, \ \forall \omega_i \in \Omega$
K2. $p(\Omega) = 1$,
K3. $p(\{\omega_i, \omega_j\}) = p(\{\omega_i\}) + p(\{\omega_j\}), \ i \neq j$.

The elements ω_i of Ω are called *elementary probability events* or simply *elementary events*.

Definition 1 is the finite version of Kolmogorov's standard definition of probability measure. In the usual definition, Ω can be an infinite set, and then $\mathcal{F}$ needs to be a σ-algebra. However, for the purpose of this article, we will restrict our discussions to finite sets Ω. For that reason, we will also refer to p as a *proper probability distribution or joint probability distribution*.

It is a well know fact that for some systems it is not possible to define a proper probability distribution. This is, in fact, the heart of Bell's inequalities showing that for an EPR-Bohm type experiment no local hidden-variable theories exist that are compatible with quantum mechanics [Bell, 1964; Bell, 1966]. In fact, a hidden variable exists if and only if a joint probability distribution exists [Suppes and Zanotti, 1981; Fine, 1982], and Bell's inequalities are a necessary and sufficient condition for the existence of a joint probability distribution [Fine, 1982; Suppes et al., 1996a].

To overcome this difficulty, the use of upper probabilities, where K3 is modified to include subadditivity, has been proposed [Suppes and Zanotti, 1991; de Barros and Suppes, 2001; Hartmann and Suppes, 2010]. The main reason for such a proposal is that quantum mechanics, as suggested by Feynman's remarks, is nonmonotonic, and upper probabilities offer a framework where nonmonotonicity can be described mathematically. Thus, before we define negative probabilities, it is useful to start with the more well-known theory of upper probabilities.

Definition 2. Let Ω be a finite set, $\mathcal{F}$ an algebra over Ω, and p^* a real-valued functions, $p^* : \mathcal{F} \to \mathbb{R}$. Then $(\Omega, \mathcal{F}, p^*)$ is an upper-probability space, and p^* an upper-probability measure, if and only if:

U1. $\quad 0 \leq p^*(\{\omega_i\}) \leq 1, \ \forall \omega_i \in \Omega$
U2. $\quad p^*(\Omega) = 1,$
U3. $\quad p^*(\{\omega_i, \omega_j\}) \leq p^*(\{\omega_i\}) + p^*(\{\omega_j\}), \ i \neq j.$

Remark 3. The main difference between upper and proper probabilities is the substitution K3 for U3.

Remark 4. In many systems of interest, where the probabilities of elementary events are computed from a set of given marginal probabilities, the inequalities from U3 imply an underdetermination for the possible values of the joint upper probability distribution for all $\omega_i \in \Omega$.

Remark 5. If follows from K2 and K3 that

$$\sum_{\omega_i \in \Omega} p(\{\omega_i\}) = 1,$$

but because of U3 is
$$\sum_{\omega_i \in \Omega} p^* (\{\omega_i\}) \geq 1.$$

One of the main difficulties with upper probabilities is that, because it uses subadditivity, it is very hard in practice to compute it. Subaditivity also implies that a large number of different upper measures exist, even when all moments are given. An usual approach is to request p^* to be as close as possible to a proper measure by minimizing the value of $\sum_{\omega_i \in \Omega} p^* (\{\omega_i\})$, which can be greater than one when no proper joint exists [Suppes and Zanotti, 1981; Suppes and Zanotti, 1991; Fine, 1994; de Barros and Suppes, 2001; Hartmann and Suppes, 2010].

Another possible approach was proposed by [Feynman, 1987] in connection to the two slit experiment: negative probabilities. Though Feynman could not find any use for negative probability, recent research has shown that there may be some advantage for using them [Abramsky and Brandenburger, 2011; Al-Safi and Short, 2013; Zhu et al., 2013; de Barros, 2014; Oas et al., 2014; Abramsky and Brandenburger, 2014; de Barros and Oas, 2014; de Barros et al., 2014]. To define negative probabilities, we first need to set forth a description of certain systems where no proper probability distribution exists. This is the goal of the following definition.

Definition 6. Let Ω be a finite set, $\mathcal{F}$ an algebra over Ω, and let $(\Omega_i, \mathcal{F}_i, p_i)$, $i = 1, \ldots, n$, a set of n probability spaces, $\mathcal{F}_i \subseteq \mathcal{F}$ and $\Omega_i \subseteq \Omega$. Then $(\Omega, \mathcal{F}, p)$, where p is a real-valued function, $p : \mathcal{F} \to [0, 1]$, $p(\Omega) = 1$, is *compatible* with the probabilities p_i's if and only if

$$\forall (x \in \mathcal{F}_i) (p_i(x) = p(x)).$$

Furthermore, the marginals p_i are *viable* if and only if p is a probability measure.

Remark 7. Intuitively, we can think of the p_i's as observable marginals. The definition above says that such marginals are *viable* if it is possible to sew them together to produce a larger probability function over the whole space Ω (in the same spirit of [Dzhafarov and Kujala, 2013a; Dzhafarov and Kujala, 2014a; de Barros et al., 2014]). Our definition is an extension of [Halliwell and Yearsley, 2013], as we consider not only the case where viable distributions exist, but also when they do not.

For some experimental situations, such as the EPR-Bell setup, the marginals are not viable, but are compatible with a p that has the characteristic of being negative for some elements of Ω (but not negative for the observable marginals). This motivates the following definition of a p that may take negative values.

Definition 8. Let Ω be a finite set, $\mathcal{F}$ an algebra over Ω, P and P' real-valued functions, $P: \mathcal{F} \to \mathbb{R}$, $P': \mathcal{F} \to \mathbb{R}$, and let $(\Omega_i, \mathcal{F}_i, p_i)$, $i = 1, \ldots, n$, a set of n probability spaces, $\mathcal{F}_i \subset \mathcal{F}$ and $\Omega_i \subseteq \Omega$. Then $(\Omega, \mathcal{F}, P)$ is a negative probability space, and P a negative probability, if and only if $(\Omega, \mathcal{F}, P)$ is compatible with the probabilities p_i's and

N1. $\quad \forall (P') \left(\sum_{\omega_i \in \Omega} |P(\{\omega_i\})| \leq \sum_{\omega_i \in \Omega} |P'(\{\omega_i\})| \right)$

N2. $\quad \sum_{\omega_i \in \Omega} P(\{\omega_i\}) = 1$

N3. $\quad P(\{\omega_i, \omega_j\}) = P(\{\omega_i\}) + P(\{\omega_j\}), \ i \neq j.$

In the above definition, we replace axiom K1 of nonnegativity with a minimization of the L1 norm of the function P. Intuitively, as with uppers, we seek a quasi-probability distribution that is as close to a proper distribution as possible. Furthermore, the departure from such proper distributions, which would have no negative numbers, motivates the following definition of M^* as a measure of this departure. Throughout this paper we use p for proper probability measures (Definition 1), p^* for upper and lower probabilities (Definition 2), and P for negative probabilities (Definition 8).

Definition 9. Let $(\Omega, \mathcal{F}, P)$ be a negative probability space. Then, the *minimum L1 probability norm*, denoted M^*, or simply *minimum probability norm*, is given by $M^* = \sum_{\omega_i \in \Omega} |P(\{\omega_i\})|$.

Proposition 10. Let $(\Omega, \mathcal{F}, P)$ be a negative probability space and $(\Omega, \mathcal{F}, p)$ a (Kolmogorov) probability space. Then $p = P$ iff $M^* = \sum p_i$.

Proof. Let us start with $M^* = \sum p_i$. It follows from it that all elementary events satisfy the condition $0 \leq P(\{\omega_i\}) \leq 1$, which is K1. Together with N2 and N3, then P is also a probability measure, and $(\Omega, \mathcal{F}, P)$ a probability space. Now, if $(\Omega, \mathcal{F}, p)$ is a probability space, it follows from K1 that $M^* = \sum_{\omega_i \in \Omega} |P(\{\omega_i\})| = \sum p_i$. ∎

Remark 11. Proposition 10 tells us that axioms N1–N3 include, as a special case, K1–K3. In other words, in the special case when a proper Kolmogorovian distribution exists ($M^* = \sum p_i$), P coincides with p.

We end this section with one last definition that is relevant to physical systems.

Definition 12. Let Ω be a finite set, $\mathcal{F}$ an algebra over Ω, and let $(\Omega_i, \mathcal{F}_i, p_i)$, $i = 1, \ldots, n$, a collection of n probability spaces, $\mathcal{F}_i \subseteq \mathcal{F}$ and $\Omega_i \subseteq \Omega$. Then the probabilities p_i are *contextually biased*[3] if there exists an a in $\mathcal{F}_i$ and in $\mathcal{F}_j$, $i \neq j$, $b \neq a \neq b'$, $\sum_{\forall b \in \mathcal{F}_j} p(a \cap b) \neq \sum_{\forall b' \in \mathcal{F}_i} p(a \cap b')$.

Remark 13. In physics, for multipartite systems, this definition is equivalent to the no-signaling condition.

Proposition 14. Let Ω be a finite set, $\mathcal{F}$ an algebra over Ω, and let $(\Omega_i, \mathcal{F}_i, P_i)$, $i = 1, \ldots, n$, a set of n probability spaces, $\mathcal{F}_i \subseteq \mathcal{F}$ and $\Omega_i \subseteq \Omega$. The probabilities P_i are not *contextuality biased* if and only if there exists a negative probability $(\Omega, \mathcal{F}, P)$ compatible with the p_i's.

Proof. See [Al-Safi and Short, 2013; Oas *et al.*, 2014; Abramsky and Brandenburger, 2014] for different proofs. ∎

3.1 Interpretations of Negative Probabilities

As we mentioned above, both Dirac and Feynman saw negative probabilities as computational devices. Though we can take such pragmatic view, as negative probabilities help explore certain situations of interest in quantum mechanics (see [Oas *et al.*, 2014] for an example), the question still remains as to their meaning. Here we discuss some proposals on how to interpret negative probabilities.

Let us start with the interpretation of negative probabilities in terms of two disjoint probability measures, μ^+ and μ^-, initially suggested by [Burgin, 2010; Burgin and Meissner, 2012] and then expanded and formalized by [Abramsky and Brandenburger, 2014]. Here we follow the interpretation as presented by [Abramsky and Brandenburger, 2014]. The main idea of this interpretation comes from the well-known fact (see [Rao and Rao, 1983]) that it is possible to decompose a signed measure μ into two non-negative ones, μ^- and μ^+, such that

$$\mu = \mu^+ - \mu^-.$$

Following this idea, [Abramsky and Brandenburger, 2014] creates two copies of the sample space, namely $\Omega \times \{+, -\}$, giving them a new dimension corresponding to $+$ or $-$. For example, for the case of three random variables **X**, **Y**, and **Z**, the set of all elementary events would be $\{\omega_{xyz}, \omega_{xy\bar{z}}, \ldots, \omega_{\bar{x}\bar{y}z}, \omega_{\bar{x}\bar{y}\bar{z}}\}$, whereas the expanded set $\Omega \times \{+, -\}$ would have as elementary events

---
[3] Here we adopt and adapt the terminology of [Dzhafarov and Kujala, 2014a].

$\{\omega_{xyz+},\ldots,\omega_{\overline{xyz}+},\omega_{xyz-},\ldots,\omega_{\overline{xyz}-}\}$. Because of the above decomposition, they define a probability over the set $+$ and $-$ such that

$$P = p^+ - p^-,$$

where now P is the negative probability and p^+ and p^- can be interpreted as proper probability distributions over $(\Omega, \pm)$. To interpret P, [Abramsky and Brandenburger, 2014] proposes an effect akin to interference. When we observe an event, say corresponding to the element ω_{xyz+}, we use p^+ as the distribution to create our data table, and similarly for ω_{xyz-}. However, the counting due to a $-$ element can annihilate a counting for a $+$ element, and vice versa. In a certain sense, this interpretation of negative probabilities is conceptually similar to some hidden variable approaches in the literature, as for example the virtual photon model of Suppes and de Barros [Suppes et al., 1996c; Suppes and de Barros, 1994a; Suppes et al., 1996b; Suppes et al., 1996d; Suppes and de Barros, 1994b; Suppes et al., 1996a] or the ESR model [Garola and Sozzo, 2009; Sozzo and Garola, 2010; Garola and Sozzo, 2011a; Garola and Sozzo, 2011b; Garola et al., 2014], to cite a few. In these approaches, an underlying hidden process can erase an outcome that would be possible if it were not for the interference of non-observable events. However, the problem with this interpretation is that, even though it is based on a frequentist view, it does not provide a way of counting actual observable clicks on a measurement device and interpret them as negative probabilities; in other words, it assumes some non-accessible reality.

We now turn to another frequentist interpretation of negative probabilities, this one proposed by Khrennikov [Khrennikov, 1993a; Khrennikov, 1993b; Khrennikov, 1994b; Khrennikov, 1994a; Khrennikov, 2009]. Khrennikov starts with the idea that, in the frequentist view, the probability of an event is defined as by the number of times such an event occurs in an infinite sequence of possible outcomes or ensembles. Following [Khrennikov, 2009], let S_N be a sequence of N ensembles with, $S_N = \{s_1, s_2, \ldots, s_N\}$. For each of the ensembles s_i, one can ask whether the property represented by the random variable $\mathbf{A}$ has the value a or not, and let $S(\mathbf{A} = a)$ be the subset of all ensembles such that $\mathbf{A} = a$. Then, in the standard frequentist interpretation, the probability $p(\mathbf{A} = a)$ is given by

$$p(\mathbf{A} = a) = \lim_{N \to \infty} \frac{|S(\mathbf{A} = a) \cap S_N|}{|S|}, \tag{1}$$

where $|\cdot|$ represents the cardinality of a set. Khrennikov then argues that there are ensembles for which the limit in (1) does not converge, and for such cases negative probabilities can be obtained as the result of a regularization procedure or order. In such sense, negative probabilities come as the result of quasi-random sequences that violate the principle of statistical stabilization [Khrennikov, 1993a]. Khrennikov then proposes to generalize probabilities coming from sequences that violate the principle of statistical stabilization as measures taking values not only on the field $\mathbb{R}$, but also on the p-adic extensions of the set of rationals $\mathbb{Q}$, i.e. $\mathbb{Q}_p$. We recall that $\mathbb{R}$ is defined, through Cauchy sequences, as the completion of $\mathbb{Q}$ under the Euclidian norm. Similarly, $\mathbb{Q}_p$ is the completion of $\mathbb{Q}$ under the p-adic norm (see [Khrennikov, 2009] for a clear exposition of $\mathbb{Q}_p$ and its properties). Once he does that, he shows that certain sequences that have probability zero in the sense of (1) would have negative probabilities in their p-adic extension, whereas sequences of probability one would have p-adic values greater than one. Thus, according to Khrennikov, we can interpret negative (greater than one) probabilities as events of probability zero (one) for sequences that violate the principle of statistical stabilization. Here we note that contextual random variables are quasi-random, and violate the principle of statistical stabilization.

We now turn to the meaning of the minimum L1 norm we propose for negative probabilities. Similarly to negative probabilities, the sub or super additivity of upper and lower probabilities allows for a large number of solutions to the joint probability that is consistent with the marginals. One possibility is to think of upper and lowers as subjective measures of belief based on inconsistent information [Suppes and Zanotti, 1991]. As such, it can be argued that, since upper and lowers do not add to one as standard probabilities do, one should choose among the many different distributions those whose sums are as close to one as possible. This is, in a certain sense, similar to what the minimum L1 norm does for negative probabilities. As such, this norm, which quantifies how much a negative probability deviates from a proper probability, provides us a measure of how inconsistent the correlations between random variables are [de Barros, 2014].

We end this section with one last general comment. Instead of using negative probabilities, it is possible to simply extend the probability space such that when we talk about correlations between experimentally observable variables, as proposed by [Dzhafarov and Kujala, 2012; Dzhafarov and Kujala, 2013a]. To understand this point, imagine we start with three variables **X**, **Y**, and **Z**, as in the above example. Instead of thinking of them as three variables, we could think of them as six, one for each experimental context: $\mathbf{X}_Y$, $\mathbf{X}_Z$, ..., $\mathbf{Z}_Y$. It is easy to show that in some important physical examples, such as the fa-

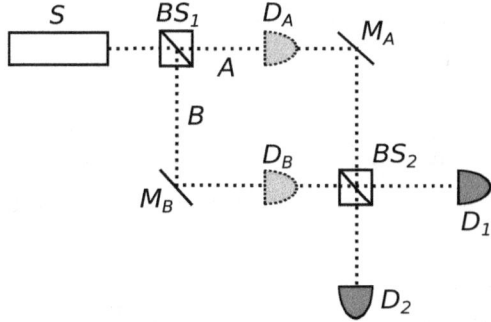

Figure 4. Schematics of a Mach-Zehnder interferometer. A light beam from a source S is divided into two equal-intensity beams by the beamsplitter BS_1. The beams are reflected by mirrors M_A and M_B, and then recombined by the beamsplitter BS_2. Photons from S are then detected by photodetectors D_1 or D_2. The count rates on D_1 and D_2 depend on the geometry of the system, in particular the optical distances between BS_1, BS_2, and M_A and M_B. In the which-path version of this experiment, detectors D_A and/or D_B may be placed on each arm of the interferometer to determine the trajectory of the photon.

mous Bell-EPR setup, such extension of the probability space is sufficient to grant the existence of a joint probability distribution, but at the cost of having $\mathbf{X}_Y \neq \mathbf{X}_Z$. Thus, the apparent inconsistencies mentioned in the previous paragraph could be argued to come from an identity assumption for the random variables: that a random variable remains the same in different contexts [Dzhafarov and Kujala, 2013b; Dzhafarov and Kujala, 2013a; Dzhafarov and Kujala, 2014a; Dzhafarov and Kujala, 2014b]. As such, the minimum norm could be interpreted as a measure of contextuality [de Barros et al., 2014; de Barros and Oas, 2014; Oas et al., 2014].

4 The Mach-Zehnder Interferometer and Negative Probabilities

Now that we saw the basic relationships between negative probabilities and upper and Kolmogorovian probabilities, we turn our attention to the two slit experiment its simplified version of the Mach-Zehnder interferometer, schematically shown in Figure 4. We will not attempt to give a full quantum-mechanical description of this experiment, but instead focus on an elementary representation of the main ideas behind it. Intuitively, each arm of the interferometer, A and B, corresponds to a possible path the particle can take, which is equivalent to a particle going through one slit or the other on the two slit experiment. However, the Mach-Zehnder differs from the double-slit experiment as the particle only has two possible outcomes, a detection in D_1 or D_2, as opposed to an infinite number of points on a screen. Such two possible outcomes could be thought as two points on the screen on the screen corresponding to a maximum and minimum intensities in the interference pattern.

To elaborate on the analogy with the two slit experiment, let us think about the experiment in terms of particles. First, we can select an interferometer setup such that we have constructive interference in D_1 and destructive in D_2. In other words, the lengths of the interferometer arms are chosen such that $P(\mathbf{D}_1 = 1) = 1$ and $P(\mathbf{D}_2 = 1) = 0$, where $\mathbf{D}_1$ ($\mathbf{D}_2$) is a ± 1-valued random variable representing a detection on D_1 (D_2) when its value is 1 and no detection when -1. From now on we will use the standard notation $p_{d_1} = P(\mathbf{D}_1 = 1)$, $p_{\bar{d}_1} = P(\mathbf{D}_1 = -1)$, $p_{\bar{d}_2} = P(\mathbf{D}_2 = -1)$ and so on. With this notation, our interferometer is such that $p_{d_1} = p_{\bar{d}_2} = 1$ and $p_{\bar{d}_1} = p_{d_2} = 0$.

Now that the interferometer is set up, let us examine the two possible classical models (according to Feynman) behind it: the wave and particle models. We start with the wave point of view. Let $\psi = A \cos(\omega t)$ represent a coherent wave arriving at the beam splitter BS_1 at time t and being split in both directions, A and B. The wave going through A is unchanged by BS_1, and arrives at M_A as $\frac{A}{2} \cos(\omega t + \phi_1)$, where ϕ_1 is a phase that depends on the geometry of the interferometer, specifically on the distance between BS_1 and M_A. At M_A it becomes $-\frac{A}{2} \sin(\omega t + \phi_1)$ due to a $\pi/2$ phase shift upon reflection, and arrives at BS_2 as $-\frac{A}{2} \sin(\omega t + \phi_1 + \phi_2)$. For the wave going through B, it arrives at M_B as $-\frac{A}{2} \sin(\omega t + \phi_2)$ and at BS_2 as $-\frac{A}{2} \cos(\omega t + \phi_2 + \phi_1)$, where we assume for the geometry that distance between BS_1 and M_B is the same as the distance between M_A and BS_2 (and similarly for BS_1 to M_A and M_B to BS_2). The beam splitter BS_2 now recombines the two waves coming from A and B, and the outputs on D_1 and D_2 are the superposition of those waves. In other words,

$$\begin{aligned}\psi_{D_1} &= -\frac{A}{2}\sin\left(\omega t + \phi_1 + \phi_2 + \frac{\pi}{2}\right) - \frac{A}{2}\cos(\omega t + \phi_2 + \phi_1) \quad (2)\\ &= -\frac{A}{2}\cos(\omega t + \phi) - \frac{A}{2}\cos(\omega t + \phi)\\ &= -A\cos(\omega t + \phi),\end{aligned}$$

where the first term on the rhs is the reflected wave from A, and $\phi =$

$\phi_1 + \phi_2$. For D_2 we obtain, with now the wave B getting a phase of $\pi/2$,

$$\begin{align}
\psi_{D_2} &= -\frac{A}{2}\sin(\omega t + \phi) - \frac{A}{2}\cos\left(\omega t + \phi + \frac{\pi}{2}\right) \tag{3}\\
&= -\frac{A}{2}\sin(\omega t + \phi) + \frac{A}{2}\sin(\omega t + \phi) = 0. \tag{4}
\end{align}$$

We can now compute the mean intensity of the entering wave, ψ,

$$I_S = \langle \psi^2 \rangle_t = \frac{A^2}{2},$$

where

$$\langle f \rangle_t = \frac{1}{T}\lim_{\omega T \gg 1} \int_t^{t+T} f\, dt'$$

represents the time average. The intensity at D_1, is

$$I_{D_1} = \frac{A^2}{2},$$

whereas the intensity at D_2 is

$$I_{D_2} = 0,$$

consistent with the value for the source S. These particular values for the intensities at D_1 and D_2 present the highest contrast between the intensities at each detector, or, as it is often referred, maximum visibility (of interference). So, to summarize, according to the wave model we see no wave energy arriving at D_2 because of destructive interference due to the relative phases of the different paths the wave traveled.

Now let us examine the view that photons are particles, and let us assume that we can control the intensity of the source such that one particle at a time goes through the interferometer. A particle comes out of the source S and enters the interferometer through the beam splitter BS_1. Beam splitters divide beams into two equal intensity ones. This translates into a particle having probability $1/2$ of going to either arm A or B. For the sake of argument, let us assume that the particle went into arm A. Once it leaves the interferometer, it is reflected at mirror M_A and reaches another beam splitter BS_2. Once again, it has probability $1/2$ of going on either direction, since it interacts only locally with BS_2. In other words, it cannot possibly have any information about the geometry of path B, or even if it is not simply closed with the presence of a physical barrier. Therefore, the probability of this particle reaching D_1 is the same as D_2, and it equals $1/2$. The same analysis can be applied to the photon going through arm B. Therefore, from a particle point of view, $p_{d_1} = p_{d_2} = 1/2$. This is in stark contradiction with the wave result.

In the standard interpretation of quantum mechanics, this contradiction is resolved by stating that one cannot simultaneously assign two complimentary properties to a quantum system. In the above case, we cannot assign the property of going through path A or B (which is what happens if we have a particle). To be able to say that a particle went through A or B, we need to actually place a detector D_A and D_B in the paths. At the same time, if we place such detectors, we destroy the wave-like behavior, and its associated probabilities at D_1 and D_2. If the detectors simply destroy the particle, then we have obviously an impossibility in obtaining the joint probability distribution in a trivial way, as we can show by the following simple example (which we also spell out in more detail below).

Let D_X and D_Y be two detectors that absorb photons, and let us put D_Y at the end of a source S that produces photons. So, if a photon is emitted, the probability of observing it is $p_{d_y} = 1$. However, if we put D_X in between S and D_Y, we will have that $p_{d_x} = 1$ and $p_{d_y} = 0$. The observable terms are the following.

$$p_{d_x d_y} = p_{\bar{d}_x \bar{d}_y} = p_{\bar{d}_x d_y} = 0,$$

$$p_{d_x \bar{d}_y} = 1.$$

But this leads to a contradiction, as

$$p_{d_y} = 1 = p_{d_x d_y} + p_{\bar{d}_x d_y} = 0.$$

That contradiction comes from an obvious reason: we have different experiments, and therefore the random variable $\mathbf{D}_Y$ representing a measurement in one experiment cannot be the same as the $\mathbf{D}_Y$ in the other experiment. The assumption of the existence of a joint distribution is equivalent to the assumption that both $\mathbf{D}_Y$'s are the same.

In the Mach-Zehnder, an analogous case to the example above would be the following. With the setup in Figure 4, we split the experiment into two types: destructive and non-destructive measurements. A destructive measurement happens when the observed system is not available for any other measurements afterward. For example, in many photodetection apparatuses, the photon is absorbed by the device and ceases to exist. A non-destructive measurement is the one where the system is available for later measurements. For each type of experiment, there are four possible experimental conditions, which we label as Case 1 to Case 8. We start with destructive measurements.

Case 1 (D_1, D_2 only) This case corresponds to the standard Mach-Zehnder with no *which-path* information, since no detector is put on either arm of the interferometer. Thus, a joint probability distribution exists for all the random variables involved. When this is the case, we have that $p_{d_1\bar{d}_2} = 1$ and $p_{d_1 d_2} = p_{\bar{d}_1 d_2} = p_{\bar{d}_1 \bar{d}_2} = 0$. Here $p_{d_1 d_2}$ and $p_{\bar{d}_1 \bar{d}_2}$ are set to zero almost by definition, as we are considering cases where we have one and only one photodetection.

Case 2 (D_1, D_2, D_A) In this case, if we have a detection on D_A, we have no detection on D_1 and D_2 (intuitively, the photon was absorbed by the detector). On the other hand, if we have no detection on D_A, D_1 and D_2 are equiprobable, since the interference effects are destroyed by the presence of a detection. Thus, $p_{d_a d_1 d_2} = p_{d_a d_1 \bar{d}_2} = p_{d_a \bar{d}_1 \bar{d}_2} = p_{d_a \bar{d}_1 d_2} = p_{\bar{d}_a d_1 d_2} = p_{\bar{d}_a \bar{d}_1 \bar{d}_2} = 0$, and $p_{\bar{d}_a d_1 \bar{d}_2} = p_{\bar{d}_a \bar{d}_1 d_2} = \frac{1}{2}$.

Case 3 (D_1, D_2, D_B) Similarly to Case 2, here $p_{d_b d_1 d_2} = p_{d_b d_1 \bar{d}_2} = p_{d_b \bar{d}_1 \bar{d}_2} = p_{d_b \bar{d}_1 d_2} = p_{\bar{d}_b d_1 d_2} = p_{\bar{d}_b \bar{d}_1 \bar{d}_2} = 0$, and $p_{\bar{d}_b d_1 \bar{d}_2} = p_{\bar{d}_b \bar{d}_1 d_2} = \frac{1}{2}$.

Case 4 ($D_1, D_2, D_A,$ and D_B) This simply tells us that we can only observe in A or B, but nowhere else, since the detectors in A and B destroy the photon, not allowing it to reach D_1 or D_2. Then $p_{d_a d_b d_1 d_2} = p_{d_a d_b d_1 \bar{d}_2} = p_{d_a d_b \bar{d}_1 d_2} = p_{d_a d_b \bar{d}_1 \bar{d}_2} = 0$, $p_{\bar{d}_a d_b d_1 d_2} = p_{\bar{d}_a d_b \bar{d}_1 d_2} = p_{\bar{d}_a d_b \bar{d}_1 \bar{d}_2} = 0$, $p_{d_a \bar{d}_b d_1 d_2} = p_{d_a \bar{d}_b \bar{d}_1 d_2} = p_{d_a \bar{d}_b \bar{d}_1 \bar{d}_2} = 0$, $p_{\bar{d}_a \bar{d}_b d_1 d_2} = p_{\bar{d}_a \bar{d}_b d_1 \bar{d}_2} = p_{\bar{d}_a \bar{d}_b \bar{d}_1 d_2} = p_{\bar{d}_a \bar{d}_b \bar{d}_1 \bar{d}_2} = 0$, and $p_{d_a \bar{d}_b d_1 \bar{d}_2} = p_{\bar{d}_a d_b \bar{d}_1 d_2} = \frac{1}{2}$.

It is easy to see that we have inconsistencies between the random variables for each case, because Case 4 gives us a joint probability distribution for all observables that is inconsistent with Case 1. This simply tells us that each experimental context gives different distributions to the random variables, as the marginal expectations are different for each experimental condition.

For the non-destructive measurement we have the following experimental outcomes.

Case 5 (D_1, D_2 only) This is clearly identical to Case 1, where $p_{d_1 \bar{d}_2} = 1$ and $p_{d_1 d_2} = p_{\bar{d}_1 d_2} = p_{\bar{d}_1 \bar{d}_2} = 0$.

Case 6 (D_1, D_2, D_A) In this case, if we have a detection on D_A, but there will also be a detection on either D_1 or D_2. Furthermore, regardless of the outcomes on D_A, detections on D_1 and D_2 are equiprobable, since the interference effects are destroyed by the presence of a detection. Thus, $p_{d_a d_1 d_2} = p_{d_a \bar{d}_1 \bar{d}_2} = p_{\bar{d}_a d_1 d_2} = p_{\bar{d}_a \bar{d}_1 \bar{d}_2} = 0$, and $p_{d_a d_1 \bar{d}_2} = p_{d_a \bar{d}_1 d_2} = p_{\bar{d}_a d_1 \bar{d}_2} = p_{\bar{d}_a \bar{d}_1 d_2} = \frac{1}{4}$.

Case 7 (D_1, D_2, D_B) Similarly to Case 2, here $p_{d_b d_1 d_2} = p_{d_b \bar{d}_1 \bar{d}_2} = p_{\bar{d}_b d_1 d_2} = p_{\bar{d}_b \bar{d}_1 \bar{d}_2} = 0$, and $p_{d_b d_1 \bar{d}_2} = p_{d_b \bar{d}_1 d_2} = p_{\bar{d}_b d_1 \bar{d}_2} = p_{\bar{d}_b \bar{d}_1 d_2} = \frac{1}{4}$.

Case 8 (D_1, D_2, D_A, and D_B) This simply tells us that we can only observe in A or B, but nowhere else, since detectors absorb the photon in A and B, not letting it reach D_1 or D_2. Then $p_{d_a d_b d_1 d_2} = p_{d_a d_b \bar{d}_1 d_2} = p_{d_a d_b d_1 \bar{d}_2} = p_{d_a d_b \bar{d}_1 \bar{d}_2} = p_{\bar{d}_a d_b d_1 d_2} = p_{d_a \bar{d}_b d_1 d_2} = 0$, $p_{\bar{d}_a \bar{d}_b d_1 d_2} = p_{\bar{d}_a \bar{d}_b \bar{d}_1 d_2} = p_{\bar{d}_a \bar{d}_b d_1 \bar{d}_2} = 0$, $p_{\bar{d}_a \bar{d}_b \bar{d}_1 \bar{d}_2} = p_{\bar{d}_a d_b \bar{d}_1 \bar{d}_2} = p_{d_a \bar{d}_b \bar{d}_1 \bar{d}_2} = 0$, and $p_{\bar{d}_a d_b \bar{d}_1 d_2} = p_{\bar{d}_a d_b d_1 \bar{d}_2} = p_{d_a \bar{d}_b d_1 \bar{d}_2} = p_{d_a \bar{d}_b \bar{d}_1 d_2} = \frac{1}{4}$.

As with the destructive measurements, we have inconsistencies between the two complementary experimental conditions. This shows that which-path information creates a context that is different from the one leading to interference. In other words, the joint probabilities obtained in the non-destructive measurement are once again incompatible with the marginals for the interference patterns contained in Case 5.

In both types of experiments, described by Cases 1 through 8, the incompatibility of contexts is reflected in the non-existence of a joint (quasi) negative probability distribution for all possible outcomes. This reflects the strong contextuality of each setup, interference or which-path, leading to observables D_1 and D_2 that are contextuality biased. It is interesting at this point to notice that this would represent, in a trivial way, a case where an experimenter could choose to observe or not D_A or D_B, and such observation would change the probabilities in D_1 and D_2. Thus, in a trivial sense, the observation of, say, D_A or not could be used to *signal* another experimenter at D_1. Though this is *not* what is usually called signaling in the literature, as it does not involve any spacelike separations between a transmitter and a receiver, it does clarify the relationship between the absence of contextual bias and the no-signalling condition. To distinguish this, [Kofler and Brukner, 2013] coined the term *signaling in time*, but here we use the term *contextual measurement biases* suggested by [Dzhafarov and Kujala, 2014a]. See [Oas et al., 2014; Dzhafarov and Kujala, 2014a] for a somewhat more detailed discussion of this point, including the relationship between the existence of probability distributions (including negative ones) and signaling.

Case 5-8 are equivalent to the spirit of Feynman's discussions about the double-slit in his 1987 paper, and has been experimentally realized by [Scully et al., 1994]. Even though [Scully et al., 1994] had access to the outcomes of Case 6, 7, and 8 to infer the joint probability distribution, they used counterfactual reasoning to compute a negative probability distribution that was consistent with Case 5. To do so, they had to discard certain measurements from their marginals, say, by only looking at cases where no detection happened at detector B in case 7. Let us examine the details of this counterfactual computations. First one needs to determine what are the actual observable conditions that constrain the marginal distributions. If we put detectors on both paths, we observe

$$P(d_a d_b) = 0, \tag{5}$$

$$P(\overline{d_a} d_b) = \frac{1}{2}, \tag{6}$$

$$P(d_a \overline{d_b}) = \frac{1}{2}, \tag{7}$$

and

$$P(\overline{d_a d_b}) = 0, \tag{8}$$

which corresponds to having only one photon at a time.

Whenever we observe in detector D_1 we do not observe in D_2, and vice versa. Furthermore, since we have a single photon, we never observe in both detectors or in neither. Finally, interference requires that we only observe in D_1. Therefore,

$$P(d_1 \overline{d_2}) = 1 \tag{9}$$

$$P(\overline{d_1} d_2) = P(d_1 d_2) = P(\overline{d_1 d_2}) = 0. \tag{10}$$

Now for what Feynman considered the disturbing issue. If we put a detector in arm A or B, from (5)-(8) we can "infer" that whenever we observe the particle *not* being in A, then the particle must be (probability 1) in B. But when we block the path, the probabilities are

$$P(\overline{d_a} d_1 \overline{d_2}) = P(\overline{d_a} d_1 d_2) = \frac{1}{2}, \tag{11}$$

and

$$P(\overline{d_b} d_1 \overline{d_2}) = P(\overline{d_b} d_1 d_2) = \frac{1}{2}. \tag{12}$$

The "disturbing" aspect comes from the nonmonotonicity of the above probabilities. How can $P(\overline{d_1} d_2) = 0$, according to 10, while $P(\overline{d_a} d_1 d_2) = \frac{1}{2}$, from (11), given that $\overline{d_a} d_1 d_2$ is a proper subset of all events where $\overline{d_1} d_2$?

Of course, this nonmonotonic property cannot be reproduced by Kolmogorov's axioms. To see this, let S_1 and S_2 be two sets in $\mathcal{F}$ such that $S_1 \subseteq S_2$ (as is the case for $S_1 = \{\omega_i \in \Omega | \mathbf{D}_1 = -1, \mathbf{D}_2 = 1\}$ and $S_2 = \{\omega_i \in \Omega | \mathbf{A} = -1, \mathbf{D}_1 = -1, \mathbf{D}_2 = 1\}$ above). Then, we can construct a set $S_1' = S_2 \setminus S_1$ such that $S_1 \cup S_1' = S_2$ and $S_1 \cap S_1' = \emptyset$. From K3 we have that $p(S_1 \cup S_1') = p(S_1) + p(S_1') = p(S_2)$, and from K1 we have at once that $p(S_1) \leq p(S_2)$ if $S_1 \subseteq S_2$, which is clearly violated by the probabilities above. Notice that in order to prove monotonicity, we had to use the non-negativity axiom K1. However, since negative probabilities violate K1, they may be nonmonotonic. For instance, from $P(S_1 \cup S_1') = P(S_1) + P(S_1') = p(S_2)$, it is possible to have $P(S_1) > P(S_2)$ if $P(S_1') < 0$.

Before we compute the joint (quasi) negative probabilities from the assumptions above, let us examine in more detail (11) and (12). The fact that each add to one corresponds to a selection of experiments where no detection happens on D_A or D_B. In other words, we are only looking at a subset of all possible experimental outcomes (essentially, this is equivalent to a postselection of data). In fact, we can see that (11) and (12) are distinct from what one observes in Case 2 and Case 3 or Case 6 and Case 7, which, as we pointed out earlier, are incompatible with Case 1 or Case 5, respectively. In this restricted data set, the counterfactual reasoning leads to a weaker context-dependency between variables, allowing for the existence of a joint negative probability distribution, as we now show.

From (11) and (12) we obtain the following set of linear equations

$$P\left(\overline{d_a} \cdot d_1 \overline{d}_2\right) = P\left(\overline{d_a} d_b d_1 \overline{d}_2\right) + P\left(\overline{d_a} \overline{d_b} d_1 \overline{d}_2\right) = \frac{1}{2}, \quad (13)$$

$$P\left(\overline{d_a} \cdot \overline{d}_1 d_2\right) = P\left(\overline{d_a} d_b \overline{d}_1 d_2\right) + P\left(\overline{d_a} \overline{d_b} \overline{d}_1 d_2\right) = \frac{1}{2}, \quad (14)$$

$$P\left(\cdot \overline{d_b} d_1 d_2\right) = P\left(d_a \overline{d_b} d_1 d_2\right) + P\left(\overline{d_a} \overline{d_b} d_1 d_2\right) = \frac{1}{2}, \quad (15)$$

and

$$P\left(\cdot \overline{d_b} d_1 \overline{d}_2\right) = P\left(d_a \overline{d_b} d_1 \overline{d}_2\right) + P\left(\overline{d_a} \overline{d_b} d_1 \overline{d}_2\right) = \frac{1}{2}. \quad (16)$$

From (9)–(10), we also obtain that

$$P(\cdot \cdot d_1 d_2) = P(d_a d_b d_1 d_2) + P\left(\overline{d_a} d_b d_1 d_2\right)$$
$$+ P\left(d_a \overline{d_b} d_1 d_2\right) + P\left(\overline{d_a} \overline{d_b} d_1 d_2\right) = 0, \quad (17)$$

$$P\left(\cdot\cdot d_1\bar{d}_2\right) = P\left(d_a d_b d_1 \bar{d}_2\right) + P\left(\bar{d}_a d_b d_1 \bar{d}_2\right)$$
$$+ P\left(d_a \bar{d}_b d_1 \bar{d}_2\right) + P\left(\overline{d_a d_b} d_1 \bar{d}_2\right) = 1, \quad (18)$$

$$P\left(\cdot\cdot \bar{d}_1 d_2\right) = P\left(d_a d_b \bar{d}_1 d_2\right) + P\left(\bar{d}_a d_b \bar{d}_1 d_2\right)$$
$$+ P\left(d_a \bar{d}_b \bar{d}_1 d_2\right) + P\left(\overline{d_a d_b} \bar{d}_1 d_2\right) = 0, \quad (19)$$

$$P\left(\cdot\cdot \bar{d}_1 \bar{d}_2\right) = P\left(d_a d_b \bar{d}_1 \bar{d}_2\right) + P\left(\bar{d}_a d_b \bar{d}_1 \bar{d}_2\right)$$
$$+ P\left(d_a \bar{d}_b \bar{d}_1 \bar{d}_2\right) + P\left(\overline{d_a d_b} \bar{d}_1 \bar{d}_2\right) = 0. \quad (20)$$

Finally, (5)–(8) yields

$$P\left(d_a d_b \cdot\cdot\right) = P\left(d_a d_b d_1 d_2\right) + P\left(d_a d_b \bar{d}_1 d_2\right)$$
$$+ P\left(d_a d_b d_1 \bar{d}_2\right) + P\left(d_a d_b \bar{d}_1 \bar{d}_2\right) = 0, \quad (21)$$

$$P\left(d_a \bar{d}_b \cdot\cdot\right) = P\left(d_a \bar{d}_b d_1 d_2\right) + P\left(d_a \bar{d}_b \bar{d}_1 d_2\right)$$
$$+ P\left(d_a \bar{d}_b d_1 \bar{d}_2\right) + P\left(d_a \bar{d}_b \bar{d}_1 \bar{d}_2\right) = \frac{1}{2}, \quad (22)$$

$$P\left(\bar{d}_a d_b \cdot\cdot\right) = P\left(\bar{d}_a d_b d_1 d_2\right) + P\left(\bar{d}_a d_b \bar{d}_1 d_2\right)$$
$$+ P\left(\bar{d}_a d_b d_1 \bar{d}_2\right) + P\left(\bar{d}_a d_b \bar{d}_1 \bar{d}_2\right) = \frac{1}{2}, \quad (23)$$

$$P\left(\overline{d_a d_b} \cdot\cdot\right) = P\left(\overline{d_a d_b} d_1 d_2\right) + P\left(\overline{d_a d_b} \bar{d}_1 d_2\right)$$
$$+ P\left(\overline{d_a d_b} d_1 \bar{d}_2\right) + P\left(\overline{d_a d_b} \bar{d}_1 \bar{d}_2\right) = 0. \quad (24)$$

As a last condition, from axiom N2, all probabilities of elementary events must add to one, i.e.,

$$P\left(d_a d_b d_1 d_2\right) + P\left(d_a d_b d_1 \bar{d}_2\right) + P\left(d_a d_b \bar{d}_1 d_2\right) + P\left(d_a d_b \bar{d}_1 \bar{d}_2\right) + \quad (25)$$
$$P\left(d_a \overline{d_b} d_1 d_2\right) + P\left(a \overline{d_b} d_1 \bar{d}_2\right) + P\left(a \overline{d_b} \bar{d}_1 d_2\right) + P\left(a \overline{d_b} \bar{d}_1 \bar{d}_2\right) +$$
$$P\left(\overline{d_a} d_b d_1 d_2\right) + P\left(\overline{d_a} d_b d_1 \bar{d}_2\right) + P\left(\overline{d_a} d_b \bar{d}_1 d_2\right) + P\left(\overline{d_a} d_b \bar{d}_1 \bar{d}_2\right) +$$
$$P\left(\overline{d_a d_b} d_1 d_2\right) + P\left(\overline{d_a d_b} d_1 \bar{d}_2\right) + P\left(\overline{d_a d_b} \bar{d}_1 d_2\right) + P\left(\overline{d_a d_b} \bar{d}_1 \bar{d}_2\right) = 1$$

The general solution to the underdetermined (and not independent) system of equations (13)–(25) is

$$P(d_a d_b d_1 d_2) = \alpha, \tag{26}$$

$$P\left(d_a d_b d_1 \bar{d}_2\right) = \theta + \frac{1}{2}(\delta - \gamma + \beta - \alpha), \tag{27}$$

$$P\left(d_a d_b \bar{d}_1 d_2\right) = -\frac{1}{2} - \theta, \tag{28}$$

$$P\left(d_a d_b \bar{d}_1 \bar{d}_2\right) = \frac{1}{2} + \frac{1}{2}(-\delta + \gamma - \beta - \alpha), \tag{29}$$

$$P\left(d_a \bar{d}_b d_1 d_2\right) = \frac{1}{2}(-\delta - \gamma + \beta - \alpha), \tag{30}$$

$$P\left(d_a \bar{d}_b d_1 \bar{d}_2\right) = \frac{1}{2} - \theta + \frac{1}{2}(-\delta + \gamma - \beta + \alpha) \tag{31}$$

$$P\left(d_a \bar{d}_b \bar{d}_1 d_2\right) = \theta, \tag{32}$$

$$P\left(d_a \bar{d}_b \bar{d}_1 \bar{d}_2\right) = \delta, \tag{33}$$

$$P\left(\bar{d}_a d_b d_1 d_2\right) = \frac{1}{2}(\delta - \gamma - \beta - \alpha), \tag{34}$$

$$P\left(\bar{d}_a d_b d_1 \bar{d}_2\right) = \frac{1}{2} - \theta + \frac{1}{2}(-\delta + \gamma - \beta + \alpha), \tag{35}$$

$$P\left(\bar{d}_a d_b \bar{d}_1 d_2\right) = \theta, \tag{36}$$

$$P\left(\bar{d}_a d_b \bar{d}_1 \bar{d}_2\right) = \beta, \tag{37}$$

$$P\left(\bar{d}_a \bar{d}_b d_1 d_2\right) = \gamma, \tag{38}$$

$$P\left(\bar{d}_a \bar{d}_b d_1 \bar{d}_2\right) = \theta + \frac{1}{2}(\delta - \gamma + \beta - \alpha), \tag{39}$$

$$P\left(\bar{d}_a \bar{d}_b \bar{d}_1 d_2\right) = \frac{1}{2} - \theta, \tag{40}$$

$$P\left(\bar{d}_a \bar{d}_b \bar{d}_1 \bar{d}_2\right) = -\frac{1}{2} + \frac{1}{2}(-\delta - \gamma - \beta + \alpha), \tag{41}$$

where α, β, γ, δ, and θ are arbitrary constants. It is clear from (41) that no nonnegative solution exists for (13)–(25). Furthermore, because the system is underdetermined, there are an infinite number of solutions that satisfy (13)–(25). To find the negative probabilities, though, we need to minimize the L1 norm, M^*. Doing so for (41) is straightforward but tedious, and we can show that such minimum happens when $0 \leq \alpha \leq \frac{1}{2}$, $\beta = 0$, $\delta = 0$, $\theta = 0$, and $\alpha = -\gamma$. This gives us the general solution minimizing M^* as

$$P(d_a d_b d_1 d_2) = \alpha, \quad P\left(d_a d_b d_1 \bar{d}_2\right) = 0, \tag{42}$$
$$P\left(d_a d_b \bar{d}_1 d_2\right) = -\tfrac{1}{2}, \quad P\left(d_a d_b \bar{d}_1 \bar{d}_2\right) = \tfrac{1}{2} - \alpha,$$
$$P\left(d_a \bar{d}_b d_1 d_2\right) = 0, \quad P\left(d_a \bar{d}_b d_1 \bar{d}_2\right) = \tfrac{1}{2},$$
$$P\left(d_a \bar{d}_b \bar{d}_1 d_2\right) = 0, \quad P\left(d_a \bar{d}_b \bar{d}_1 \bar{d}_2\right) = 0,$$
$$P\left(\bar{d}_a d_b d_1 d_2\right) = 0, \quad P\left(\bar{d}_a d_b d_1 \bar{d}_2\right) = \tfrac{1}{2},$$
$$P\left(\bar{d}_a d_b \bar{d}_1 d_2\right) = 0, \quad P\left(\bar{d}_a d_b \bar{d}_1 \bar{d}_2\right) = 0,$$
$$P\left(\overline{d_a d_b} d_1 d_2\right) = -\alpha, \quad P\left(\overline{d_a d_b} d_1 \bar{d}_2\right) = 0,$$
$$P\left(\overline{d_a d_b} \bar{d}_1 d_2\right) = \tfrac{1}{2}, \quad P\left(\overline{d_a b d_b} \bar{d}_1 \bar{d}_2\right) = -\tfrac{1}{2} + \alpha,$$

$0 \le \alpha \le \tfrac{1}{2}$, clearly showing that $M^* = 3$. We should notice that this value of M^* is greater than the $M^* = 2$ for the Bell-EPR case [Oas et al., 2014], perhaps already suggesting that the double-slit is more contextual (see [de Barros et al., 2014] for a discussion of M^* as a measure of contextuality). This stronger contextuality probably comes from the use of triple moments in the Mach-Zehnder as opposed to only pairwise two-moments in the case of the standard Bell-EPR setup.

Now that we have a negative probability distribution, we can use it to compute conditional probabilities based on the previous counterfactual assumptions. For instance, a standard question is this: if a photon is detected on D_1, what is the probability that this photon went through A and B? Using

$$P(d_a|d_1) = \frac{1}{N}\left[P(d_a d_b d_1 d_2) + P\left(d_a \bar{d}_b d_1 d_2\right) + P\left(d_a d_b d_1 \bar{d}_2\right) + P\left(d_a \bar{d}_a d_1 \bar{d}_2\right)\right],$$

where

$$N = P(d_a d_b d_1 d_2) + P\left(\bar{d}_a d_b d_1 d_2\right) + P\left(d_a \bar{d}_b d_1 d_2\right) + P\left(d_a d_b d_1 \bar{d}_2\right)$$
$$+ P\left(\overline{d_a d_b} d_1 d_2\right) + P\left(\bar{d}_a d_b d_1 \bar{d}_2\right) + P\left(d_a \bar{d}_b d_1 \bar{d}_2\right) + P\left(\overline{d_a d_b} d_1 \bar{d}_2\right),$$

and (42) we have that

$$P(d_a|d_1) = \frac{1}{2} + \alpha,$$

and

$$\frac{1}{2} \le P(d_a|d_1) \le 1.$$

Similarly,

$$P(d_b|d_1) = \frac{1}{N}\left[P\left(d_a d_b d_1 d_2\right) + P\left(\overline{d_a} d_b d_1 d_2\right) \right.$$
$$\left. + P\left(d_a d_b d_1 \overline{d_2}\right) + P\left(\overline{d_a} d_b d_1 \overline{d_2}\right)\right],$$

where

$$N = P\left(d_a d_b d d_2\right) + P\left(d_a \overline{d_b} d_1 d_2\right) + P\left(\overline{d_a} d_b d_1 d_2\right) + P\left(d_a d_b d_1 \overline{d_2}\right)$$
$$+ P\left(\overline{d_a} \overline{d_b} d_1 d_2\right) + P\left(d_a \overline{d_b} d_1 \overline{d_2}\right) + P\left(\overline{d_a} d_b d \overline{d_2}\right) + P\left(\overline{d_a} \overline{d_b} d_1 \overline{d_2}\right),$$

and we get

$$P(d_b|d_1) = \frac{1}{2} + \alpha,$$

the same value we got for the conditional $P(d_b|d_1)$. For d_2, the conditional probability is not defined, as the probability for d_2 from the joint is zero. However, if we set the interferometer such that the probability of d_2 is not zero, but close to it, then $P(b|d_2)$ can be shown to approach $P(d_b|d_2) = -\frac{1}{2} + \alpha$. If that is the case, it is reasonable to assume that $\alpha = 1/2$, such that we do not have negative probabilities for d_b (conditioned on d_2). If we do so, we reach the interesting conclusion that both d_b and d_a have probability 1 given an observation on d_1. In other words, if we use the counterfactual reasoning from negative probabilities, we reach the conclusion, as Feynman often said, that the particle goes through *both* paths simultaneously.

We now end this section with a discussion of some well-known uses of counterfactual reasoning in quantum mechanics and their relationship to our discussion above. First, it is worth mentioning that the famous Leggett and Garg [Leggett and Garg, 1985] setup can be thought of as similar to our double-slit experiment [Kofler and Brukner, 2013]. To see this, we recall that in Leggett and Garg (LG), measurements in three distinct times can coded by three ±1-valued random variables, say **X**, **Y**, and **Z** [Bacciagaluppi, 2014; Dzhafarov and Kujala, 2014a; de Barros et al., 2014; Dzhafarov and Kujala, 2014c]. In an analogy with the double-slit, and following [Kofler and Brukner, 2013], we can think of **X** as a measurement of position before BS_1, **Y** as a measurement of which path (say, with **Y** = 1 corresponding to A and **Y** = −1 to B), and **Z** corresponding to a detection in either D_1 (for **Z** = 1) or D_2 (**Z** = −1). This case would correspond to contextual bias, and would not include counterfactual reasoning. However, in the original LG paper, counterfactual reasoning happens by not measuring **Y**, but instead making inferences about **Y** from an absence of detection in one of the paths. As we mentioned above, such contextual bias is not surprising, as the effect of measuring could be thought as interfering with the experimental conditions themselves. This is similar to what happens in our analysis above.

Something analogous happens with the argument given by [Scully et al., 1994]. In his paper, he talks about negative probabilities, and shows that they lead to the interference between two possible modes. However, it is easy to see that the negative probabilities so obtained are only existent because of the same type of counterfactual reasoning shown above. That should be clear by the fact that, in their experiment, the interference pattern is existent, and therefore we have observational contextual bias, similar to the Leggett-Garg setup.

5 Final Remarks

In this paper we presented a proposed theory of negative probabilities that could be used to describe non-monotonic reasoning. Such theory was shown to be equivalent, in the case when a proper probability distribution exists, to the standard Kolmogorov probability, as the requirement of minimizing the total probability mass leads to a Kolmogorovian distribution. Furthermore, in cases where no proper joint probability exists, the minimization of the total negative mass is simply a requirement that our quasi-probability distribution is as close to a proper one as possible. Such minimization is similar to the requirement with upper probabilities of minimizing the total sum of probabilities, which may exceed one.

We did not attempt to interpret negative probabilities, but instead took the approach that they constitute a bookkeeping tool that meets a minimum rationality criteria of minimization of the L1 probability norm. This criteria is perhaps not without practical consequences. For instance, in [de Barros, 2014], we showed that in certain cases where the pairwise correlations lead to contradictions, this minimization results in constraints to the triple moments. In addition, in [Oas et al., 2014], it was shown that the minimized L1 norm is equivalent to the CHSH parameter, S, as used in measures of non-locality [Cirel'son, 1980], specifically $M^* = S/2$. Finally, the L1 norm can be thought of as a measure of contextuality for random variables, and is closely related to other measures of contextuality, at least for three and four random-variables, and for more variables it suggests possible different classifications for contextuality [de Barros et al., 2014].

Though the double-slit experiment is the archetypical in discussions of how quantum mechanics leads to violation of the laws of probability or logic [Dalla Chiara and Giuntini, 2014], it is not the simplest and most accessible example that contains the key conceptual elements relevant to the subject. In fact, the Mach-Zehnder interferometer, as we showed above, presents the same characteristics as the double-slit experiment that are relevant to conceptual discussions of quantum mechanics, without the complications associated with the details of continuous interference patterns present in the double-slit. For that reason, in our discussion of the double-slit experiment in terms of negative probabilities, we resorted to the simpler case of the Mach-Zehnder interferometer. As we saw, the Mach-Zehnder interferometer allows us to talk about the features of double-slit experiment in terms of discrete random variables, which tremendously reduce the mathematical complexity without loosing any conceptual generality.

As we showed, the two possible setups for the Mach-Zehnder interferometer, one with which-path information and another with interference, present contextual biases. This has, as a consequence, the non-existence of a joint negative (quasi) probability distribution consistent with all observations of the two Mach-Zehnder interferometer setups. This clearly corresponds to two different experimental contexts, and not only does a joint probability consistent with both contexts not exist, but no negative joint distribution exists either. This type of system, where contextuality comes from contextual biases, exhibit what one could think of as stronger contextuality than other systems, such as EPR.

This stronger contextuality may be what is reflected in the large values of M^* for the two slits, but such a connection has not been studied in detail. In fact, it is interesting to notice that the large values of M^* is associated to a set of observables that do not provide a complete picture of the experimental conditions, as it relies on counterfactuals. Perhaps there is a connection between large M^* for a restricted set of observables and contextual biases for an extended set, which could provide an interesting criteria for contextual biases. Notice that, as mentioned in Section 4, contextual biases are equivalent to the violation of the no-signaling condition in multipartite systems.

Finally, we would like to comment on Feynman's [Feynman, 1987] discussion of the double-slit experiment. In this paper, he argues that the non-monotonic character of quantum probabilities could be represented by non-observable negative probabilities. He then goes on and constructs (in a very informal way) a possible negative probability that could explain the outcomes of the experiment. However, as we pointed out above, negative probabilities consistent with the outcomes of the double-slit experiment are impossible, unless we make use of certain specific counterfactual reasoning. It is interesting to note that in his actual experimental realization of Feynman's double-slit experiment, Scully et al. [Scully *et al.*, 1994] construct a negative probability distribution, and their probabilities rely exactly on the type of counterfactual reasoning used above. Were they to try and construct a negative probability from the full range of experimental data, they would not be able to do so. A similar case is present in the LG experiment also discussed above.

Acknowledgements Two of the authors (JdB, GO) would like to express their gratitude for having the honor to contribute to this volume recognizing Patrick Suppes. It is with great sadness that Pat's passing came so soon. He did not have a chance to contribute to this final version, but the core ideas put forth here stem from him, and we believe that he would be pleased with the final result. However, we emphasize that any errors are the exclusive responsibility of the first two authors, and would not be present if the paper had gone through Pat's usual rigorous review. We also want to take this opportunity to express our indebtedness to Pat for his guidance and patience over the past few decades. Pat introduced both of us to the importance of joint probability distributions in quantum mechanics, and was to us not only a collaborator and mentor, but also a friend, and we heartily dedicate this paper to him.

This paper stems from a 2011 seminar led by the three authors, where a full review and exploration of negative probabilities was undertaken, and we profited from exchanges with the seminar participants, in particular Tom Ryckman, Claudio Carvalhaes, Michael Heaney, Michael Friedman, Hyungrok Kim, and Niklas Damiris. It was Pat Suppes's goal to establish a rigorous, consistent theory of negative probability, and this work is in line with that unfinished goal. We also benefited tremendously from collaborations and discussions with Ehtibar Dzhafarov and Janne Kujala. We also thank Guido Bacciagaluppi for discussions and for pointing us to the Kofler and Brukner reference.

BIBLIOGRAPHY

[Abramsky and Brandenburger, 2011] S. Abramsky and A. Brandenburger. The sheaf-theoretic structure of non-locality and contextuality. *New Journal of Physics*, 13(11):113036, November 2011.

[Abramsky and Brandenburger, 2014] S. Abramsky and A. Brandenburger. An operational interpretation of negative probabilities and no-signalling models. In F. van Breugel, E. Kashefi, C. Palamidessi, and J. Rutten, editors, *Horizons of the Mind. A Tribute to Prakash Panangaden*, number 8464 in Lecture Notes in Computer Science, pages 59–75. Springer Int. Pub., 2014.

[Al-Safi and Short, 2013] S.W. Al-Safi and A.J. Short. Simulating all nonsignaling correlations via classical or quantum theory with negative probabilities. *Physical Review Letters*, 111(17):170403, October 2013.

[Aspect et al., 1981] Alain Aspect, Philippe Grangier, and Gérard Roger. Experimental tests of realistic local theories via bell's theorem. *Physical Review Letters*, 47(7):460–463, August 1981.

[Aspect et al., 1982] Alain Aspect, Jean Dalibard, and Gérard Roger. Experimental test of bell's inequalities using time- varying analyzers. *Physical Review Letters*, 49(25):1804–1807, December 1982.

[Bacciagaluppi, 2014] G. Bacciagaluppi. Leggett-garg inequalities, pilot waves and contextuality. *International Journal of Quantum Foundations*, 1(1):1–17, 2014. arXiv: 1409.4104.

[Bell, 1964] J.S. Bell. On the einstein-podolsky-rosen paradox. *Physics*, 1(3):195–200, 1964.

[Bell, 1966] J.S. Bell. On the problem of hidden variables in quantum mechanics. *Rev. Mod. Phys.*, 38(3):447–452, 1966.

[Burgin and Meissner, 2012] Mark Burgin and Gunter Meissner. Negative probabilities in financial modeling. *Wilmott*, 2012(58):60–65, 2012.

[Burgin, 2010] Mark Burgin. Interpretations of negative probabilities. *arXiv:1008.1287 [physics, physics:quant-ph]*, August 2010. arXiv: 1008.1287.

[Cirel'son, 1980] B.S. Cirel'son. Quantum generalizations of bell's inequality. *Letters in Mathematical Physics*, 4(2):93–100, March 1980.

[Dalla Chiara and Giuntini, 2014] M.L. Dalla Chiara and R. Giuntini. The two-slits experiment, quantum logic and quantum computational logic. In V. Fano, editor, *Gino Tarozzi Philosopher of Physics. Studies in the philosophy of entanglement on his 60th birthday*. Franco Angeli, Milan, Italy, 2014.

[de Barros and Oas, 2014] J. Acacio de Barros and Gary Oas. Negative probabilities and counterfactual reasoning in quantum cognition. *Physica Scripta*, 2014(T163):014008 (6pp), April 2014.

[de Barros and Suppes, 2000] J. Acacio de Barros and P. Suppes. Inequalities for dealing with detector inefficiencies in greenberger-horne-zeilinger type experiments. *Physical Review Letters*, 84(5):793–797, 2000.

[de Barros and Suppes, 2001] J. Acacio de Barros and P. Suppes. Probabilistic results for six detectors in a three-particle GHZ experiment. In Jean Bricmont, D. Durr, Maria Carla Galavotti, G. Ghirardi, F. Petruccione, and N. Zanghi, editors, *Chance in Physics*, volume 574 of *Lectures Notes in Physics*, page 213, 2001.

[de Barros et al., 2014] J. Acacio de Barros, E.N. Dzhafarov, J.V. Kujala, and G. Oas. Unifying two methods of measuring quantum contextuality. *arXiv:1406.3088 [quant-ph]*, June 2014. arXiv: 1406.3088.

[de Barros, 2014] J. Acacio de Barros. Decision making for inconsistent expert judgments using negative probabilities. In H. Atmanspacher, E. Haven, K. Kitto, and D. Raine, editors, *Quantum Interaction Quantum Interaction*, Lecture Notes in Computer Science, pages 257–269. Springer, Berlin/Heidelberg, 2014.

[Dzhafarov and Kujala, 2012] E.N. Dzhafarov and J.V. Kujala. Selectivity in probabilistic causality: Where psychology runs into quantum physics. *Journal of Mathematical Psychology*, 56(1):54–63, February 2012.

[Dzhafarov and Kujala, 2013a] E.N. Dzhafarov and J.V. Kujala. All-possible-couplings approach to measuring probabilistic context. *PLoS ONE*, 8(5):e61712, May 2013.

[Dzhafarov and Kujala, 2013b] E.N. Dzhafarov and J.V. Kujala. Random variables recorded under mutually exclusive conditions: Contextuality-by-default. *arXiv:1309.0962 [quant-ph]*, September 2013.

[Dzhafarov and Kujala, 2014a] Ehtibar N. Dzhafarov and Janne V. Kujala. Contextuality in generalized klyachko-type, bell-type, and leggett-garg-type systems. *arXiv:1411.2244 [physics, physics:quant-ph]*, November 2014. arXiv: 1411.2244.

[Dzhafarov and Kujala, 2014b] E.N. Dzhafarov and J.N. Kujala. A qualified kolmogorovian account of probabilistic contextuality. *Lecture Notes in Computer Science*, 8369:201–212, 2014.

[Dzhafarov and Kujala, 2014c] E.N. Dzhafarov and J.V. Kujala. Generalizing bell-type and leggett-garg-type inequalities to systems with signaling. *arXiv:1407.2886 [quant-ph]*, July 2014.

[Dzhafarov and Kujala, 2014d] E.N. Dzhafarov and J.V. Kujala. Probabilistic contextuality in EPR/Bohm-type systems with signaling allowed. *arXiv:1406.0243 [quant-ph, q-bio]*, June 2014.

[Einstein et al., 1935] A. Einstein, B. Podolsky, and N. Rosen. Can quantum-mechanical description of physical reality be considered complete? *Physical Review*, 47(10):777–780, May 1935.

[Feynman et al., 2011] R.P. Feynman, R.B. Leighton, and M.L. Sands. *The Feynman lectures on physics: Mainly mechanics, radiation, and heat*, volume 3. Basic Books, New York, NY, 2011.

[Feynman, 1987] R.P. Feynman. Negative probability. In B.J. Hiley and F.D. Peat, editors, *Quantum implications: essays in honour of David Bohm*, pages 235–248. Routledge, London and New York, 1987.

[Fine, 1982] A. Fine. Hidden variables, joint probability, and the bell inequalities. *Physical Review Letters*, 48(5):291–295, February 1982.

[Fine, 1994] Terrence L. Fine. Upper and lower probability. In Paul Humphreys, editor, *Patrick Suppes: Scientific Philosopher*, number 234 in Synthese Library, pages 109–133. Springer Netherlands, January 1994.

[Garola and Sozzo, 2009] C. Garola and S. Sozzo. The ESR model: A proposal for a noncontextual and local hilbert space extension of QM. *EPL (Europhysics Letters)*, 86(2):20009, April 2009.

[Garola and Sozzo, 2011a] Claudio Garola and Sandro Sozzo. Generalized observables, bell's inequalities and mixtures in the ESR model for QM. *Foundations of Physics*, 41(3):424–449, March 2011.

[Garola and Sozzo, 2011b] Claudio Garola and Sandro Sozzo. The modified bell inequality and its physical implications in the ESR model. *International Journal of Theoretical Physics*, 50(12):3787–3799, December 2011.

[Garola et al., 2014] C. Garola, M. Persano, J. Pykacz, and S. Sozzo. Finite local models for the GHZ experiment. *International Journal of Theoretical Physics*, 53(2):622–644, February 2014.

[Greenberger et al., 1989] D. M Greenberger, M. A Horne, and A. Zeilinger. Going beyond {B}ell's theorem. In M. Kafatos, editor, *Bell's theorem, Quantum Theory, and Conceptions of the Universe*, volume 37 of *Fundamental Theories of Physics*, pages 69–72. Kluwer, Dordrecht, Holland, 1989.

[Halliwell and Yearsley, 2013] J. J. Halliwell and J. M. Yearsley. Negative probabilities, fine's theorem, and linear positivity. *Physical Review A*, 87(2):022114, February 2013.

[Hartmann and Suppes, 2010] Stephan Hartmann and Patrick Suppes. Entanglement, upper probabilities and decoherence in quantum mechanics. In Mauricio Suárez, Mauro Dorato, and Miklós Rédei, editors, *EPSA Philosophical Issues in the Sciences*, pages 93–103. Springer Netherlands, January 2010.

[Howard et al., 2014] Mark Howard, Joel Wallman, Victor Veitch, and Joseph Emerson. Contextuality supplies the /'magic/' for quantum computation. *Nature*, 510(7505):351–355, June 2014.

[Jacques et al., 2007] Vincent Jacques, E. Wu, Frédéric Grosshans, François Treussart, Philippe Grangier, Alain Aspect, and Jean-François Roch. Experimental realization of wheeler's delayed-choice gedanken experiment. *Science*, 315(5814):966–968, February 2007.

[Khrennikov, 1993a] A. Khrennikov. p-adic probability theory and its applications. {T}he principle of statistical stabilization of frequencies. *Theoretical and Mathematical Physics*, 97(3):1340–1348, December 1993.

[Khrennikov, 1993b] A. Khrennikov. p-adic statistical models. *Doklady Akademii Nauk*, 330(3):300–304, 1993.

[Khrennikov, 1994a] A. Khrennikov. Discrete {Q}p-valued probabilities. *Russian Academy of Sciences-Doklady Mathematics-AMS Translation*, 48(3):485–490, 1994.

[Khrennikov, 1994b] A. Khrennikov. *p-Adic Valued Distributions in Mathematical Physics*, volume 309 of *Mathematics and Its Applications*. Springer Science+Business Media, Dordrecht, Holland, 1994.

[Khrennikov, 2009] A. Khrennikov. *Interpretations of probability*. Walter de Gruyter, Berlin, 2009.

[Kochen and Specker, 1967] Simon Kochen and E. P. Specker. The problem of hidden variables in quantum mechanics. *Journal of Mathematics and Mechanics*, 17:59–87, 1967.

[Kochen and Specker, 1975] S. Kochen and E. P. Specker. The problem of hidden variables in quantum mechanics. In C.F. Hooker, editor, *The Logico-Algebraic Approach to Quantum Mechanics*, pages 293–328. D. Reidel Publishing Co., Dordrecht, Holland, 1975.

[Kofler and Brukner, 2013] Johannes Kofler and Časlav Brukner. Condition for macroscopic realism beyond the leggett-garg inequalities. *Physical Review A*, 87(5):052115, May 2013.

[Kolmogorov, 1950] A.N. Kolmogorov. *Foundations of the theory of probability*, volume viii. Chelsea Publishing Co., Oxford, England, 1950.

[Leggett and Garg, 1985] A.J. Leggett and A. Garg. Quantum mechanics versus macroscopic realism: Is the flux there when nobody looks? *Physical Review Letters*, 54(9):857–860, March 1985.

[Markiewicz et al., 2013] M. Markiewicz, P. Kurzynski, J. Thompson, S.-Y. Lee, A. Soeda, T. Paterek, and D. Kaszlikowski. Unified approach to contextuality, non-locality, and temporal correlations. *arXiv:1302.3502 [quant-ph]*, February 2013.

[Oas et al., 2014] G. Oas, J. Acacio de Barros, and C. Carvalhaes. Exploring non-signalling polytopes with negative probability. *Physica Scripta*, 2014(T163):014035 (5pp), 2014.

[Rao and Rao, 1983] K. P. S. Bhaskara Rao and M. Bhaskara Rao. *Theory of Charges: A Study of Finitely Additive Measures*. Academic Press, May 1983.

[Scully and Druhl, 1982] Marlan O. Scully and Kai Druhl. Quantum eraser: A proposed photon correlation experiment concerning observation and "delayed choice" in quantum mechanics. *Physical Review A*, 25(4):2208–2213, April 1982.

[Scully et al., 1994] Marlan O. Scully, Herbert Walther, and Wolfgang Schleich. Feynman's approach to negative probability in quantum mechanics. *Physical Review A*, 49(3):1562–1566, March 1994.

[Silverman, 1995] Mark P. Silverman. *More than one mystery: explorations in quantum interference*. Springer-Verlag, New York, 1995.

[Sozzo and Garola, 2010] Sandro Sozzo and Claudio Garola. A hilbert space representation of generalized observables and measurement processes in the ESR model. *International Journal of Theoretical Physics*, 49(12):3262–3270, December 2010.

[Suppes and de Barros, 1994a] P. Suppes and J. Acacio de Barros. A random-walk approach to interference. *International Journal of Theoretical Physics*, 33(1):179–189, 1994.

[Suppes and de Barros, 1994b] Patrick Suppes and J. Acacio de Barros. Diffraction with well-defined photon trajectories: A foundational analysis. *Foundations of Physics Letters*, 7(6):501–514, December 1994.

[Suppes and Zanotti, 1981] Patrick Suppes and Mario Zanotti. When are probabilistic explanations possible? *Synthese*, 48(2):191–199, 1981.

[Suppes and Zanotti, 1991] P. Suppes and M. Zanotti. Existence of hidden variables having only upper probabilities. *Foundations of Physics*, 21(12):1479–1499, 1991.

[Suppes et al., 1996a] Patrick Suppes, J. Acacio de Barros, and Gary Oas. A collection of probabilistic hidden-variable theorems and counterexamples. In R. Pratesi and L. Ronchi, editors, *Waves, Information and Foundation of Physics: a tribute to Giuliano Toraldo di Francia on his 80th birthday*, Florence, Italy, 1996. Italian Physical Society.

[Suppes et al., 1996b] Patrick Suppes, J. Acacio de Barros, and Adonai S Sant'Anna. A proposed experiment showing that classical fields can violate bell's inequalities. *arXiv:quant-ph/9606019*, June 1996.

[Suppes et al., 1996c] Patrick Suppes, J. Acacio de Barros, and Adonai S. Sant'Anna. Violation of bell's inequalities with a local theory of photons. *Foundations of Physics Letters*, 9(6):551–560, December 1996.

[Suppes et al., 1996d] Patrick Suppes, Adonai Sant'Anna, and J. Acacio de Barros. A particle theory of the casimir effect. *Foundations of Physics Letters*, 9(3):213–223, 1996.

[Wheeler, 1978] J. A Wheeler. The "past" and the "delayed-choice" double-slit experiment. *Mathematical Foundations of Quantum Theory*, page 9, 1978.

[Zhu et al., 2013] Xuanmin Zhu, Qun Wei, Quanhui Liu, and Shengjun Wu. Negative probabilities and information gain in weak measurements. *Physics Letters A*, 377(38):2505–2509, November 2013.

Logical Reflections on the Semantic Approach

Décio Krause and Jonas R. Becker Arenhart

ABSTRACT. Patrick Suppes is one of the first and most eminent defenders of the semantic approach to scientific theories. As is well known, the main feature of the semantic approach is supposed to be the shift from the focus on languages and sentences to models. According to this view, it is models that play the most important role in representing reality. Even though this is taken to be the "standard" approach to the problem nowadays, there is little consensus on what models are, and, consequently, on what scientific theories are. A great variety of proposals have been developed in the literature, and Suppes has emphasized that, in this case, models should be understood as set-theoretical structures. This is a major step towards converting the semantic approach into a rigorous position. However, even if we choose that route, some fine logical points remain to be discussed. In this paper we propose to reflect on the role of the underlying mathematical basis employed in the formulation of the semantic approach in the Suppesian sense. Once models are taken to be set-theoretical structures, we should not expect that the underlying set theory should go unnoticed. We analyze the influence of the underlying set theory employed in the construction of the models and argue that philosophers of science would benefit from that kind of foundational study. Some common slogans of the semantic approach, such as "a theory is a class of models", "in philosophy of science we should employ mathematics and not metamathematics" and "a theory is language free", will be examined and their 'truth' will be properly assessed.

1 Introduction

It is widely held nowadays that the semantic view successfully replaced the syntactic view on what concerns the philosophical discussion on the nature of scientific theories and has become the current orthodoxy. As philosophical legend has it, the syntactic view suffered from a broad range of difficulties, going from some intricacies due to some features of the project to external (well-founded) criticism. The *locus* of the main opposition between the semantic and the syntactic approach, as it is usually put, concerns the role of language in the formulation and identification of a scientific theory: while the syntactic view focused on formal axiomatizations, the semantic view focuses on models; that is, while the first approach (it is said) took theories to be linguistic entities, the latter (it is said) considers them to be non-linguistic entities.

As the tale usually goes, the syntactic conception of scientific theories identified a theory with its linguistic axiomatic formulation. In order to present a theory, then, one has to present both a logical and a non-logical vocabulary, to state the rules of sentence formation, to choose among the sentences some of them to play the role of axioms, and finally, to go ahead and derive theorems. The relationship between such a theory and the empirical data is provided by a set of *correspondence rules* that attribute meaning (at least) to some of the theoretical terms (Suppe (1977) chap.4 E).

In some sketches of this view, philosophers have held that the logical vocabulary and the logical axioms comprise only what is presently known as first-order logic, so that the deductive apparatus would be restricted only to that logic.[1] By its very linguistic nature, the identification of a theory with a linguistic object already raises some questions which may have troubled the proponents of the syntactic view; here we shall recall only one, which is directly related to the identification mentioned. The trouble comes from the fact that by employing distinct sets of primitive logical and/or non-logical vocabulary we may present alternative axiomatizations of what one would intuitively call "the same theory". In this case, despite the fact that we are getting different axiomatizations of "the same subject", identification of a theory with its formulation would amount to counting the distinct axiomatizations as distinct theories, for they may differ on what concerns the primitive terms and postulates assumed.

On the other hand, it is generally told that the semantic view steered away from language and linguistic formulations and took models as playing the major role in the presentation of a scientific theory. To present a theory, one must somehow describe a class of models, the models of the theory (of course, that is plainly circular if taken as a precise characterization, but let us just leave that for the moment). A quick look in the philosophical literature on models is enough to grant that the word *model* is far from univocal, and to be sure one must specify first what is meant by that word. In the beginning of the model theo-

---

[1] Although Suppe acknowledges that in what he calls "the final version" of the approach the first-order languages could be enlarged by modal operators (Suppe (1977), p. 50).

retic approach in the 1960s, the precursors of the view relied on models in the same sense as the newly born Tarskian model theory: models are taken to be mathematical entities, set theoretical structures of the same kind as used in logic. Being non-linguistic,[2] characterizing a theory in terms of models, it seems, the semantic approach would not suffer from the difficulties the syntactic approach had to face. For instance, it is said that the semantic approach is much closer to actual scientific practice, since scientists are model-builders by nature, while axiomatizing and proving theorems would be usually not their concern at all.

The opposition between both views, as we have presented it, is, we agree, rather schematic (and we do not agree with everything the tradition attributes to the syntactic view). However, it is enough for us to present the main tension created by the growing self-awareness inside the semantic tradition itself that some use of language is not wholly indispensable for this approach. That is, one must use the resources of some linguistic formulation of a theory in the semantic tradition too, and that reliance on linguistic formulations may be more than a mere auxiliary step to go to the required class of models (which could then be considered "language free"). To put the point again, given that a theory is to be considered as a class of models, how do we select the relevant models? Just point to a set of sentences in some language which may have the required structures as models, that is, sentences forming an axiom system for that class of structures. That is clear from the very Tarskian tradition being followed by the semantic view. However, as some have already pointed out (see Chakravartty (2001), Muller (2011), and Thomson-Jones (2006)), that feature makes the alleged major distinctions between the semantic view and the syntactic view look rather exaggerated. Maybe some of the features of the syntactic view that were seen as deplorable by its critics' eyes may plague also the semantic view.

However, the repudiation of the syntactic view, as we mentioned before, was not only related to the essential use of a linguistic formulation, but also with some of the internal difficulties associated with the project. Here, 'the project' means the Logical Empiricists' view of science that was coupled with the syntactic approach from its very inception. Their desire to attribute empirical significance to science according to their empirical tenets required that the specific vocabulary of a theory should be divided in two parts: the observational vocabulary and the theoretical vocabulary. Observational vocabulary had

---

[2] In terms, for A. Church, in his classical book (1956) §09 has already called our attention to the fact that "there is a sense in which semantics can be reduced to syntax".

empirical significance through a somehow direct correlation with experience, while theoretical vocabulary had a more difficult and indirect association. Anyhow, this distinction and the necessity to link theoretical terms with empirical observation led the Logical Empiricists to a whole bunch of concepts that were, indeed, very far away from actual scientific practice. Probably it is their required distinction between theoretic and observational terms and their far too strict theory of meaning applied to the interpretation of scientific theories that led the syntactic view to its demise, not the use of a linguistic formulation of a theory. Of course, that does not mean that the syntactic view without this division in vocabulary and its consequences is a better alternative for philosophers of science, but rather that the main difference between the semantic and the syntactic approaches lies not in the use of some language in presenting the theory, but in how theory relates to experience. That is, it is not necessary that by employing language in some way the semantic view should be directly identified with the syntactic view: there is a specific approach to the relation between theory and experience which accounts for their main feats and differences.

In this paper we focus on the relationship of the semantic view with language, hoping to enlighten this relationship. We shall first argue that there is a historical confusion which the pioneering works of Suppes somehow avoided, and which should be reconsidered on a reasonable formulation of the semantic approach. Also, after having freed ourselves of that confusion we shall present two alternative ways of formulating a scientific theory, both pursued by the characterization of a set-theoretical predicate. The first one, which we call axiomatization in *Suppes' style*, writes the axioms in the language of set theory (*i.e* some set theory) properly,[3] while the second one, which we call *da Costa-Chuaqui style*, advances the axioms in a specific formal language (here we follow the distinction advanced in Krause, Arenhart and Moraes (2011)). Both approaches have their merits, both, as we shall see, are in complete agreement with the semantic tradition as we formulate it, and more than that, they both serve for distinct purposes for the philosopher of science. We finish by discussing the role of formal methods in the philosophy of science. Our view is that rigorous discussions backed by logical analysis should be provided for where available; increase in clarity and avoidance of pitfalls may help us gaining a deeper understanding of the semantic approach and, consequently, of the nature of

---

[3]Other frameworks could be used instead; it should be noted that Carnap (1958) made use of higher-order logic, contrary to the standard understanding on the syntactic approach. A categorical approach was also advanced by the structuralist view of Sneed and others, see for instance Mormann (1996).

scientific theories.

2 Suppes against the syntactic approach

To begin by making clear how did some philosophers arrive at the idea that the semantic approach requires that we avoid any use of language, let us begin by recalling how Suppes himself saw the opposition between the new semantic approach and the syntactic approach. We believe that by recalling some of Suppes' arguments against the syntactic tradition we may have a better idea of the underlying motivations of the semantic approach and of the problems it was supposed to solve. This may serve our purposes in establishing that it is not the case that any kind of use of language in the semantic approach should be seen as a return to the deceased syntactic view; language, properly employed, is a part of the semantic approach. Indeed, the way language is employed in the semantic approach may have important consequences for us to understand its scope and its limitations.

We first recall that the syntactic approach is identified by Suppes himself as including axiomatization within a system of first-order logic as one of its main features (which he calls *standard formalization*, or *intrinsic characterization* (Suppes (2002) p.5)). The problem, as he puts it, is whether a theory *can* be axiomatized this way and whether such an axiomatization is *useful for gaining philosophical insights into the nature of the scientific enterprize*.

Of course, the main trouble with such a framing of the approach comes from the fact that the whole of the mathematics employed by the theory being axiomatized will need to be developed from scratch too. The presentation of a highly mathematical theory like General Relativity will have a mathematical introduction which will take more space and work to be stated in the required formal framework than the whole specific aspects of the theory proper. So, it seems, introductory work will require so much effort that the main goal, the study of the nature of scientific theories will be lost in the details involved in the axiomatization process. Furthermore, that way of proceeding is so far away from actual scientific practice that one cannot hope to get a better understanding of current science by following this approach.

To overcome that kind of difficulty, Suppes suggests that it would be rather more appropriate to have an axiomatization in his sense, namely, an *extrinsic characterization*, which defines the theory's class of models directly inside some informal set theory. Since all of the required mathematics is supposed to be already available in the set theory, all we must care about is to furnish some axioms to capture the intended set of models. So, the shift to models allows us to deal only with the required structures that may be employed to characterize the theory. Furthermore, in defending the semantic approach against the syntactic one, Suppe (2000) p. S104 claims that this is exactly what we do in practice, so that the semantic approach is in tune with scientific practice. Then, the problem turns to the axiomatizability of this class of structures; as Suppes says,

> One of the simplest ways of providing such an extrinsic characterization is simply to define the intended class of models of the theory. To ask if we can axiomatize the theory is then just to ask if we can state a set of axioms such that the models of these axioms are precisely the models in the defined class. Suppes (2002), *loc.cit.*

That is, axiomatization is not to be avoided by the semantic approach, it is rather *required* by the semantic approach. The nature of the axioms, in the sense of whether they are stated in some formal language (which, for our later purposes we may take as being some first or higher order language) or directly in some informal set theory, is not an issue settled automatically by the simple fact that we adopt the semantic approach. Suppes himself is ambivalent as to which option one should adopt. In some places, Suppes speaks as if models were taken in the Tarskian sense, namely, as structures "satisfying a theory" (see Suppes (1960), p. 289). In other places (see Suppes (2011), sec. 2), however, Suppes recognizes that if by "semantic view" we mean the stipulation of a set of structures satisfying a set of sentences of a sharply designed formal language, then his view is not the semantic view, since axiomatization by a set theoretical predicate in Suppesian sense does not require satisfaction of a set of sentences in the Tarskian sense (see also Muller (2011), sec.6). Our point is that both views are allowable for the semantic approach, and constitute merely distinct versions of this view (see our discussion below).

Before we go ahead, let us pause for a minor consideration of the first alleged opposition between the syntactic and the semantic approaches. The point until now has been that a shift to the semantic approach would allow us to avoid axiomatizing the whole of mathematics required for the characterization of a scientific theory from rock bottom. Now, as we have mentioned, it is not true that the adherents of the syntactical view adopted really a first order language to advance their axiomatizations (it is enough to take a look at the examples advanced in Carnap (1958)). Indeed, they took Russell's type theory as their standard logic (with Carnap studying diverse alternative logical systems), and just as it happens in ZFC, all of the mathematics required for a scientific theory may be developed inside type theory, although in a more cumbersome way than inside ZFC. So, why couldn't they just claim as Suppes does that the math is already available, and proceed directly to the specific axioms? Despite the fact the working inside type theory is not really something we are used to, there is nothing objectionable to that move. This consideration strengthens our position, according to which it is not the use of axioms or of a particular language that distinguishes the semantic and the syntactic view, but rather further considerations which are taken into account in the next point.[4]

For a second point, we must recall that besides an axiomatization of a theory, the syntactic approach had to include some coordination rules to provide an empirical interpretation of the theory. The vocabulary had to be divided in two parts: the theoretical and the observational vocabulary. The axioms of the theory had to be formulated by employing the theoretical vocabulary, and the meanings of terms referring to unobservable items had to be attributed by coordinating rules that related them with the observational consequences expressed in the observational vocabulary. There is a bunch of problems related with the distinction of the vocabulary and the crude interpretation of observational terms.

Here comes one of Suppes' strongest claims in favor of the semantic approach. Indeed, one of the main advantages of dealing with models rather than with a linguistic formulation concerns the relationship between theory and experimental data. The raw correspondence between interpreted terms and experience based on the observational vocabulary is the main target of Suppes in many occasions (in particular, see Suppes (1962), (1967) pp. 62-63). As we have mentioned, the syntactic approach, employed mainly for the anti-metaphysical purposes of the Logical Empiricists, had a rather crude view on these relationships.

According to Suppes ((1967) *loc. cit.*, (2002)) the relationship of a theory with phenomena is not direct, for there is much elaboration and idealization to make the phenomena apt to quantitative treatment. We proceed by the elaboration of data structures, which gradually evolve to more and more elaborate structures. Theories do not face experience directly, rather they are related to data structures (see the account in Suppes (1962)). The results taken from brute data must be labored upon to the construction of a data structure, which would be shown to be iso-

---

[4]We do not deny, of course, that working inside informal set theory is much easier and in accordance to common practice. Our point is only that the idea that one must axiomatize the mathematics of the theory together with the theory is based on a misconception of the logical positivists' approach, in particular of the logical bases they employed.

morphic to some arithmetical structure in order to "attribute numbers to things", which on its turn enables us to go from qualitative information to quantitative concepts. It is data structures which are then embedded in some model of the theory. What is relevant here is that the characterization of theories in terms of models, or set-theoretical structures, is in no way problematical just because they are taken as axiomatized by some set of sentences. Indeed, Suppes requires that there is some characterization of the theory in such a way as to obtain its models (set-theoretical structures), but for some purposes it does not matter that the characterization is provided for in some logic (for instance, first-order logic), or rather inside set theory directly.

Let us make this point clear: models play a central role in characterizing theories due to their relationships with data models. The crude account of these relationships by the syntactical view is seen by Suppes as its main defect and accounts for its unrealistic and sketchy nature. It is not the use of logical notation by itself that is problematical. That is, there is nothing to prevent one from axiomatizing a theory in some formal system of first- or higher-order logic, but it should be noticed that the whole of mathematics employed by the theory will have to be axiomatized together, rendering the process tedious and artificial. In the case of highly mathematical theories of physics, for instance, the set theoretical characterization of structures is to be preferred because it avoids a whole bunch of distracting work, given that the underlying mathematics is taken as already available. In any case, however, the same set theoretical structures are present in both kinds of axiomatization.

We may quote Suppes in defending the use of models ((1962) p.252; note also the mentioned ambivalence as to the sense of "model" being employed):

> To provide complete mathematical flexibility I shall speak of theories axiomatized within general set theory by defining an appropriate set theoretical predicate (e.g. "is a group") rather than theories axiomatized directly within first-order logic as a formal language. [...] But whichever sense of formalization is used, essentially the same logical notion of model applies.

Working inside set theory spares us from a detailed treatment of the whole mathematical basis required by the theory, but is not mandatory for the view Suppes is advocating. The essential is that we must be able to provide the set-theoretical models needed to account for the relationship of the theory with the data models (which are themselves structures too). So, the first problem, axiomatization in first-order logic, is really a defect of the syntactical approach as long as it is artificial and, in most of the times, does not reflect scientific practice. However, the real problem comes when to that axiomatization one adjoins the idea that an interpretation should be given through some coordinating rules to relate theory with experience, along with a division in the vocabulary between observational and theoretical terms. That is how the second great difficulty of the syntactical approach is supposed to be solved by the focus on models.

That is how both issues are related in the Suppesian approach: axiomatization through set theoretical predicates makes it easier for us to furnish models that may fit data structures. Data and theory are related in very complex ways. The axiomatization, as we have been discussing, is an essential step in gathering the adequate structures. To present axioms one must employ one language or another, but the required structures must be somehow characterized (and according to the language employed, as we shall discuss in the next section, distinct characterizations may originate).

So, there are two ingredients involved in the discussion: the use of set theory in an axiomatization and the use of models to account for the relation with data and (indirectly) with phenomena. In the next section we investigate how axiomatization may proceed according to two distinct fronts. On one possible path, we explicit Suppes approach to gather the structures by employing a set theoretical predicate in which the language of set theory itself is employed. On the other front, a specific language is designed, and the axioms for a theory are written in that language. In this second case, although there is no escape from axiomatizing the underlying mathematics of a theory, greater rigor is achieved and some interesting results from metamathematics may be employed. Our main point, as we shall defend in the concluding section, is that distinct kinds of analysis may be provided for by both methods. Distinct kinds of insight may be achieved in any kind of approach, as also shared by the structuralist view (see Stegmuller (1979)) so that the one to be used will depend on our aims and purposes, that is, by *pragmatic* criteria. We shall propose, however, that both approaches are part of the semantic approach.

3 Structures and models

We shall be interested in presenting two basic methods of axiomatization, two approaches to characterize a class of structures. One of them is recommended by Suppes himself, and proceeds by employing the language of set theory and all of its resources in the description of the axioms. The other was advanced by da Costa and Chuaqui (see (1988)), and the axioms are presented in a conveniently designed logical vocabulary. As we shall see, both approaches differ in strength and technical resources available to them; both serve for distinct purposes accordingly inside the semantic approach. More important than that, both deserve to be classified as part of the semantic approach, and are indeed so called (sometimes confusingly) in the literature. We intend to clarify matters by distinguishing them and by presenting their main features.

3.1 Structures

We shall be working inside first-order Zermelo-Fraenkel set theory with the axiom of choice, ZFC (but of course this framework could be modified if necessary, see Krause and Arenhart (2012)). Our first immediate goal is to present a general theory of structures that should be rigorous enough for our attempt in designing both versions of the semantic approach we shall investigate in this section. We basically follow da Costa and Rodrigues (2007) in the definitions.

We begin by defining the concept of type:

Definition 3.1 (Types). The set τ of types is the least set satisfying the following conditions:

1. (i) $i \in \tau$

2. (ii) If $a_0, a_1, \ldots, a_{n-1} \in \tau$, then $\langle a_0, a_1, \ldots, a_{n-1}\rangle \in \tau$, with $1 \leq n < \omega$.

As usual, ω stands for the set of all natural numbers. Next we introduce the concept of *order* of a type in τ.

Definition 3.2 (Order). If $a \in \tau$, the order of a (henceforth $\mathrm{ord}(a)$), is defined as follows:

1. (i) $\mathrm{ord}(i) = 0$

2. (ii) $\mathrm{ord}(\langle a_0, a_1, \ldots, a_{n-1}\rangle) = \max\{\mathrm{ord}(a_0), \mathrm{ord}(a_1), \ldots, \mathrm{ord}(a_{n-1})\} + 1$.

In the next definition, the usual set-theoretical operations of power set $\mathcal{P}$ and cartesian product $\times$ are being employed.

Definition 3.3. Given a non-empty set D, we define a function t, having τ as its domain, as follows:

1. (i) $t(i) = D$

2. (ii) If $a_0, a_1, \ldots, a_{n-1} \in \tau$, then $t(\langle a_0, a_1, \ldots, a_{n-1}\rangle) = \mathcal{P}(t(a_0) \times t(a_1) \times \ldots \times t(a_{n-1}))$.

Thus, following the usual extensional set-theoretical definition of a relation as a set of ordered pairs, for each type a, the mapping t gives us the set of all relations of that type, built from the elements of the domain D. Obviously, as the orders of the types increase, the objects attributed to them by t increase in complexity too. We say that the elements of $t(a)$ have type a. For example, the elements of type i are, by definition, the elements of D, and are called *individuals*. Then, $\langle i \rangle$ is the type of the properties of individuals, $\langle i, i \rangle$ is the type of the binary properties of individuals, $\langle\langle i \rangle\rangle$ is the type of the unary properties of unary properties of individuals, that is, sets of sets of individuals, and so on.

Definition 3.4 (Scale). The *scale* based on D is the set $\bigcup(\text{range } t)$, denoted by $\varepsilon(D)$.

Definition 3.5 (Cardinal). The cardinal κ_D associated to $\varepsilon(D)$ is defined as

$$\kappa_D = \sup\{|D|, |\mathcal{P}(D)|, |\mathcal{P}^2(D)|, \ldots\}.$$

Definition 3.6 (Structure). A structure $\mathfrak{A}$ based on D is an ordered pair of the form

$$\mathfrak{A} = \langle D, R_\iota \rangle$$

where R_ι is a sequence of elements of $\varepsilon(D)$ (with ι ranging over a set of indices), and we suppose that the domain of this sequence has cardinal strictly less than κ_D. We say that κ_D is the cardinal associated with $\mathfrak{A}$, and that $\varepsilon(D)$ is the scale associated with $\mathfrak{A}$.

As follows from our definition of t and of a scale, each element of $\varepsilon(D)$ has a certain type, for it belongs to $t(a)$ for some $a \in \tau$. Now, the order of a relation is defined as the order of its type. The order of $\mathfrak{A}$, denoted $\text{ord}(\mathfrak{A})$, is the order of the greatest of the types of the relations of the family R_ι, if there is one, and if there is no such relation, we put $\text{ord}(\mathfrak{A}) = \omega$.

If a structure has order κ, we say that it is an *order*$-\kappa$ structure. Groups are examples of order-1 structures, for they contain only order-1 relations (actually, depending on the formalization, they contain just one order-1 relation of type $\langle i, i, i \rangle$, for the binary operation on the domain can be seen as a ternary relation). There are some simple examples of structures which are not order-1 structures such as well-orderings, cyclic groups, Dedekind-complete ordered fields, and topological spaces. Empirical theories, whose structures are highly abstract and employ heavy mathematics are also not order-1 structures, and this is quite important to note, for as a consequence, the models of such theories *are not* order-1 structures (typical of standard model theory) as it seems to be assumed by many philosophers.[5]

3.2 The language of a structure

Of course, our goal is to speak of structures and of all the objects of a scale. To do that in an adequate and general way, we consider two basic higher-order languages, termed (cf. da Costa and Rodrigues (2007)) $\mathcal{L}^\omega_{\omega\omega}(R)$ (or simply $\mathcal{L}^\omega(R)$) and $\mathcal{L}^\omega_{\omega\kappa}(R)$.

In general, an infinitary language $\mathcal{L}^\eta_{\mu\kappa}$, with $\kappa < \mu$ being infinite cardinals (or ordinals) and $1 \leq \eta \leq \omega$, enables us to consider conjunctions and disjunctions of $n \leq \mu$ formulas and blocks of quantifiers with $m < \kappa$ many quantifiers. The superscript η indicates the order of the language (first-order, second-order, etc.). In both cases R is the set of the non-logical constants of the language. Thus, in $\mathcal{L}^\omega_{\mu\omega}(R)$ ($\omega < \mu$) we may have infinitely many conjunctions and disjunctions of formulae, but blocks of quantifiers with finitely many quantifiers only. $\mathcal{L}^\omega_{\omega\omega}(R)$ is a higher-order language, suitable for type theory (higher-order logic). Standard first-order languages are of the kind $\mathcal{L}^1_{\omega\omega}$, and the same happens to $\mathcal{L}_\in$, the basic language of ZFC. Put in a more precise way,

Definition 3.7 (Order of a language). A language $\mathcal{L}^n_{\mu\kappa}$, with $1 \leq n < \omega$, is called a language of order n. A language $\mathcal{L}^\omega_{\mu\kappa}$ is said to be of order ω.

A language of order n contains only types variables of order $t \leq n$ and quantification over variables of types having order $\leq n-1$ (da Costa and Rodrigues (2007), p.8).

In order to exemplify how we can define a higher-order language using a first-order language (such as $\mathcal{L}_\in$), let us sketch the language $\mathcal{L}^\omega_{\omega_1\omega}(R)$, but we could consider whatever infinitary language $\mathcal{L}^\omega_{\mu\kappa}$, provided that the involved cardinals exist in ZFC (for instance, we couldn't use an inaccessible cardinal).

The primitive symbols of $\mathcal{L}^\omega_{\omega_1\omega}(R)$ are the following ones:

1. (i) Sentential connectives: $\neg, \wedge, \vee, \rightarrow, \bigwedge,$ and $\bigvee$.

2. (ii) Quantifiers: $\forall$ and $\exists$

---

[5]Indeed, it seems to us that when philosophers speak of models of scientific theories, they do not realize that these structures are in general not order-1 structures. It is common to find philosophers speaking about applying Löwenheim-Skolem or the Compactness theorems to scientific theories in the semantic approach.

3. (iii) For each type t, a family of variables of type t whose cardinal is ω.

4. (iv) Primitive relations: for any type t, a collection of constants of that type (possibly some of them may be empty). The collection of these constants form the set R. For instance, for groups, R comprises constants for S and $\circ$, while for vector spaces it comprises constants for V, F, $+$ and $\cdot$ at least.

5. (v) Parentheses: left and right parentheses ('(' and ')'), and comma (',').

6. (vi) Equality: $=_t$ of type $t = \langle t_1, t_2 \rangle$, with t_1 and t_2 of the same type.

Variables and constants of type t are *terms* of that type. If T is a term of type $\langle t_1, \ldots, t_n \rangle$ and $T_1, \ldots, T_n$ are terms of types $t_1, \ldots, t_n$ respectively, then $T(T_1, \ldots, T_n)$ is an *atomic formula*. If T_1 and T_2 are terms of the same type t, then $T_1 =_{\langle t_1, t_2 \rangle} T_2$ is an *atomic formula*. We shall write $T_1 = T_2$ for this last formula, leaving the type of the identity relation implicit. If α, β, α_i are formulas (with α_i a family of formulas such that $i \leq \omega_1$), then $\neg \alpha$, $\alpha \wedge \beta$, $\alpha \vee \beta$, $\alpha \to \beta$, $\bigwedge \alpha_i$, and $\bigvee \alpha_i$ are formulas. Then, we are able to write formulas with denumerably many conjunctions and disjunctions. Furthermore, if X is a variable of type t, then $\forall X \alpha$ and $\exists X \alpha$ are also formulas (and only finite blocks of quantifiers are allowed). These are the only formulas of the language. The concepts of free and bound variables and other syntactic concepts can be introduced as usual.

Now let $\mathfrak{A} = \langle D, R_\iota \rangle$ be a structure, where $R_\iota \in R$, that is, the primitive relations of the structure are chosen among the constants of our language $\mathcal{L}^\omega_{\omega_1 \omega}(R)$.[6] Then, $\mathcal{L}^\omega_{\omega_1 \omega}(R)$ can be taken as a language for $\mathfrak{A} = \langle D, R_\iota \rangle$, provided that $\kappa_D = \omega$ (recall that κ_D is the cardinal associated to $\mathfrak{A}$). Still working within (say) ZFC, we can define an interpretation of $\mathcal{L}^\omega_{\omega_1 \omega}(R)$ in $\mathfrak{A} = \langle D, R_\iota \rangle$ in an obvious way, so as what we mean by a sentence S of $\mathcal{L}^\omega_{\omega_1 \omega}(R)$ (a formula without free variables) being *true* in such a structure in the Tarskian sense, in symbols,

$$\mathfrak{A} \models S. \tag{1}$$

In the same vein, we can define the notion of *validity*. A sentence S is valid, and we write

$$\models S,$$

---

[6] We use the same symbol for constants of the language and set theoretical relations in the structure interpreting it, hoping that the context will be enough to distinguish them.

if $\mathfrak{A} \models S$ for every structure $\mathfrak{A}$.

Important to emphasize that we are describing the language $\mathcal{L}^{\omega}_{\omega_1\omega}(R)$ using the resources of some set theory such as ZFC. This way, we can speak of denumerable many variables, for instance, in a precise way. In this sense, any symbol of $\mathcal{L}^{\omega}_{\omega\kappa}(R)$, as we have remarked already, can be seen as a name for a set. Thus, '(' (the left parenthesis), for example, names a set (for a detailed description of formal languages inside set theory and its relation to interpretations, see Arenhart and Moraes (2013), sec. 3).

Now let $\mathfrak{A} = \langle D, R_\iota \rangle$ be a structure, while $\text{rng}(R_\iota)$ denotes the range of R_ι. Remember that R_ι stands for a sequence of relations of the scale $\varepsilon(D)$, that is, it is a mapping from a finite ordinal into a collection of relations in the scale. Thus $\text{rng}(R_\iota)$ stands for just the set of these relations. So, $\mathcal{L}^{\omega}_{\omega\omega}(\text{rng}(R_\iota))$ (= $\mathcal{L}^{\omega}(\text{rng}(R_\iota))$) is *the basic language of the structure* (it is not the only language to speak of the structure, for other stronger languages encompassing it could be used instead). In this case, we can interpret a sentence containing constants in $\mathcal{L}^{\omega}(\text{rng}(R_\iota))$ in $\mathfrak{A} = \langle D, R_\iota \rangle$, and to define the notion of truth for sentences of this language according to this structure in an obvious way.

To summarize, given a certain structure, we may construct within ZFC a formal language that "speaks" of the elements of the scale based on the domain of the structure. The above definitions are quite general, but show that we can construct very strong languages within ZFC for the purposes of talking about certain structures.

4 Da Costa-Chuaqui approach

The approach we shall call da Costa-Chuaqui (CCA for short) is a first step towards a rigorous presentation of how a certain class of structures may stand for a theory (see da Costa and Chuaqui (1988) and also Arenhart and Moraes (2013)). That is, given an informal scientific theory, we may employ the methods provided by this approach to present the theory as a class of models. We begin with this approach because it is not the original version intended by Suppes (not, at least, taking into account his latter claims; but check the ambivalence mentioned in section 2), nor by latter philosophers endorsing the semantic view. CCA really adds some rigor to the semantic view by requiring that the postulates of the theory be formulated in a precisely stated formal language, as described in the previous subsection 3.2[7].

---

[7]We do not claim that this is the only way to provide for such an axiomatization; the structuralist approach has some similar strategies, see Balzer *et. al.* (1987) and Hinst (1996)

To put the approach in a nutshell, following the original presentation in da Costa and Chuaqui (1988), to present a theory we proceed as follows:

- We start with a structure built in ZFC according to 3.1 (or any other set theory which may happen to be the framework we are working in). This structure is supposed "to structure" some particular field of knowledge we are interested in. For instance, we can suppose a typical group structure of the form $\langle G, \circ \rangle$ or the McKinsey, Sugar and Suppes' structure for a classical particle mechanics, $\mathfrak{P} = \langle P, \mathbf{s}, m, \mathbf{f}, \mathbf{g} \rangle$ (cf. Suppes (2002), p.320).

- we search for an appropriate formal language for that structure;

- in this language, we formulate the axioms which will have the structure as a model in the Tarskian standard sense of model.[8]

Note that here we try to characterize a class of models, although they are not in general order-1 structures; as a consequence, the languages employed to stipulate the axioms are in general not first-order languages. It is a characteristic of this approach that different languages may be used for this task, each of them with its costs and benefits, for there are infinitely many formal languages that could be used as suitable for some given structure, cf. da Costa and Rodrigues (2007).

For a simple example of a mathematical theory, one may consider the mathematical theory of groups. As usually stated, a group is a structure $\mathfrak{A} = \langle G, \circ \rangle$, where $G \neq \emptyset$ and the binary operation $\circ$ satisfies the usual postulates of group theory, that is (the quantifiers range on G):

G1 $\forall x \forall y \forall z (((x \circ y) \circ z) = (x \circ (y \circ z)))$

G2 $\exists z (\forall x (x \circ z = z \circ x = x) \wedge \forall x \exists y (x \circ y = y \circ x = z))$

Here we only quantify over individuals of G, but the use of a second order language can be advisable to deal with subgroups. A set theoretical predicate for group theory in da Costa-Chuaqui style is a sentence in the language of set theory, augmented by the symbols we are using

---

[8]We need to call the reader's attention to a particular exception here; suppose we think of a structure of the form $\langle V, \in \rangle$ suitable for ZFC proper. In this case, the models would not be *sets* of ZFC as we are suggesting. This difficulty can be surmounted by assuming a stronger theory in the metalevel, but we shall leave this point to be discussed at another time, given that the theories we are emphasizing here have models which are sets of ZFC.

now (in order to form the set R of the non-logical symbols of the language of group theory), of the following kind (for a detailed treatment, see Arenhart and Moraes (2013)):

Definition 4.1 (Group). A group is a structure X satisfying the following set theoretical predicate:

$$\mathfrak{G}(X) \iff \exists G \exists \circ (X = \langle G, \circ \rangle \wedge \wedge X \models G1 \wedge X \models G2)$$

So, for instance, we know that $\langle \mathbb{R}, + \rangle$ is a group because when $\circ$ is interpreted as denoting $+$ and the quantifiers range over $\mathbb{R}$, we can prove (in ZFC) that $\langle \mathbb{R}, + \rangle \models G1$ and $\langle \mathbb{R}, + \rangle \models G2$ in the Tarskian sense. For further examples and developments, see da Costa and Chuaqui (1988), Arenhart and Moraes (2013), Krause, Arenhart, and Moraes (2011). Now, one may simply think that this only makes explicit what was really intended by the semantic approach since the beginnings. However, considering the discussion in section 2, we have seen that there is an ambivalent relation between the semantical tradition and the use of language to characterize a set theoretical predicate. One of the reasons to prefer not to use formal languages in characterizing the intended class of models is that simplicity is lost. Apart from that, there are other main points advanced against the use of formal languages which may threaten the semantical approach as presented in the da Costa-Chuaqui style. Now we advance some of them and argue that they are not definitive against this version of the semantic approach.

First of all, let us consider once again the example presented above concerning group theory (we discuss group theory for the sake of simplicity). It will be easily recognizable that structures as $\mathfrak{A} = \langle G, \circ, 0, ^{-1} \rangle$ subjected to the following list of axioms are also examples of groups, once we consider the following axioms:

G'1 $\forall x \forall y \forall z (((x \circ y) \circ z) = (x \circ (y \circ z)))$

G'2 $\forall x (x \circ 0 = 0 \circ x = x)$

G'3 $\forall x (x \circ x^{-1} = x^{-1} \circ x = 0)$

The set theoretical predicate, in this case is the following one:

Definition 4.2 (Group'). A group is a structure X satisfying the following predicate: $\mathfrak{G}(X) \iff \exists G \exists \circ \exists^{-1} \exists 0 (X = \langle G, \circ, ^{-1}, 0 \rangle \wedge X \models G'1 \wedge X \models G'2 \wedge X \models G'3)$.

It is easy to notice that the structures satisfying $G1 - G2$ and $G'1 - G'3$ are distinct, for they have distinct signatures. The classes of structures collected by both predicates are distinct, despite the fact that both are conceived as the same theory. So, how can we say that the same theory is being characterized in both cases? This difficulty may be pointed as an artifact of our rigorous characterization of a set theoretical predicate in terms of a formal language, and may be seen as betraying the spirit of the semantic approach. Can we rest assured that this is enough for us to abandon CCA approach? Not really. Muller (2011) noticed the difficulty and provided a particular solution: instead of considering the models of a particular formulation of the axioms, we may provide models for all the theories formulated in intertranslatable lexicons, so that we have an as near as possible formulation of the semantic view as being language-free. The theories are the classes of all models of the distinct intertranslatable lexicons. That is supposed to provide for both a class of models insensitive to the particular formulation of the axioms while still retaining the Tarskian account of models, which requires some set of sentences which the models are said to model.

Another way of doing this without having to look for translations of languages consists in employing a very interesting tool which has not been explored by philosophers of science yet: the general notions of definability and invariance developed in da Costa and Rodrigues (2007). Those are clearly model theoretic notions, and allow for generalizations of some tools of model theory for higher order languages and structures (recall that usual model theory deals with order-1 structures only). If we consider infinitary higher order languages as the language of our structure, as we mentioned in 3.2, we shall call, following da Costa and Rodrigues (2007), a concept definable in this language as being *definable in the wide sense*. Now, two structures $\mathfrak{A}$ and $\mathfrak{B}$ are called equivalent, written $\mathfrak{A} \equiv \mathfrak{B}$ iff the primitive relations of each of them are definable in terms of the primitive relations of the other. So, we may think of two structures with distinct signatures as presenting the same theory only when they are equivalent in this sense.

Of course, this answer does not give us a language-free characterization of a theory (neither does Muller's solution mentioned above). However, perhaps the dream of a language free characterization of a theory is not achievable; CCA's approach does not hide the fact that a language must be employed to characterize a theory, and distinct languages may do the job. Notice that we do not identify a theory with its class of models. Since we may have distinct classes of models characterizing intuitively the same theory, we only claim that the set theoretical predicate "presents" a theory, and as the discussion above has shown, a theory may be presented in various ways. We shall have more to say about this relation of an informal theory, as used by scientists, with the relevant set theoretical predicate characterizing it in the concluding section. We believe that this kind of foundational work is relevant from a philosophical point of view, for it may help us in enlightening the structure of scientific theories.

Another criticism that may be posed concerns the existence of non-standard models. Indeed, according to Suppe's account of the syntactic view, once a theory is formulated in axiomatic form, the Löwenheim-Skolem theorem may be employed to provide for highly unintended models of the theory (see Suppe (2000), p. S104). In practice, however, we only pick the intended ones. Unintended models made the analysis of confirmation, explanation, etc., difficult for the syntactic approach. Is this not an objection also for the CCA approach? To discuss it, there are some points that should be made clear. First of all, the semantics in the syntactic approach was performed by the coordination rules, the somehow "direct" empirical interpretation. Models according to model theory are mathematical entities. Secondly, Löwenheim-Skolem theorem works only for order−1 structures, and most interesting scientific theories are of a higher order. As a third point, we shall see in the next section that even the Suppesian approach, in formulating a theory directly inside some set theory, may be plagued by unintended models and some further metamathematical artifacts. So, if we are going to have to handle with those issues anyway, this should not be pointed as a defect, but should rather be recognized and studied looking for some profitable results from a logico-foundational point of view.

So, we may keep the idea that models are structures in the standard Tarskian sense.[9] Thomson-Jones (2006) called such models *truth makers*, meaning that they are employed to make a set of sentences true. According to him (2006) pp.533-534, the semantic view should not consider models as truth makers, but only as informally interpreted mathematical structures representing the systems under investigation. There are two main criticisms addressed to the use of models in the above Tarskian sense. First of all, in taking into account the language characterizing the models, we distance ourselves from actual scientific practice. And, as we mentioned before, one of the main troubles of the syntactic approach is precisely its artificiality and distance from scientific practice. The second criticism concerns the use of set theory to characterize the structures that will be the models. Models in the Tarskian approach are set theoretical structures; employing such structures we must trust set theory and confine ourselves to the tools and limits of the particular set theory in question. Such an approach, however, is claimed to be distracting when it comes to deal with the major con-

---

[9] With the exception we have made before concerning theories like ZFC.

cerns for the philosophers of science: themes such as the relation between theory and phenomena, confirmation of theories, and scientific explanation. Does the use of a particular set theory have any relevance in discussing these issues? What if we use a radically different set theory such as Rosser-Quine's NF system? No answer to this point has been, as far as we know, taken into account till today.

Our answer to the second point above will bear also on the first one. According to our view, when it comes to the study of the foundations of science one must be as explicit as possible relatively to the logical basis and all kinds of presuppositions. So, to make clear which set theory (or, more generally, which mathematics) is being employed is a requirement for a successful foundational study. More than that, the argument may be turned on its head. As Suppes himself suggested (see Suppes (2011)), the choice of an adequate formal apparatus may be relevant in getting closer to actual scientific practice. Suppes *loc. cit.* suggests that constructive mathematics should be employed in the study of data structures to reflect the finite character of experimental practice. That is, the structures used to characterize mature scientific theories use in general lots of tools of mathematics (continuous functions, infinite quantities of all kinds), while the structures reflecting actual experiments are discrete and finitistic (see also Suppes (1967) p. 63). So, the idea that a definite mathematical framework limits our investigation may be reverted to the benefit of the philosopher of science in both directions: clearing foundational issues and keeping in touch with actual experimental practice.

Of course, this brings to light our answer to the first objection. The idea that we may become closer to actual experimentation (and, consequently, to scientific practice) by the use of some distinct mathematical tool reveals that the use of a specific language and logical apparatus is not in disagreement with the impetus of the semantic approach to be in touch with what scientists do. By its very nature, the study of the foundations of science requires some tools; it is a kind of second-order investigation which may be seen as describing and investigating more or less properly an activity performed by scientists. The claim that restriction to set theoretical structures is a limitation gets things wrong, in our view. Every approach to models that allows for mathematical structures will have to make clear which is the underlying framework, otherwise we fall in vagueness and hand-waving discourse. Issues that interest philosophers of science may be more properly addressed once those topics are clearly settled.

For one example of the productive interplay between foundational study in the semantic approach and scientific practice, we may mention that the use of set theoretical structures inside some clearly formulated set theory helps in making a very important philosophical issue clearer, namely, the complex relation between data and theory. To recall very briefly, we mention the fact that to go from qualitative data to quantitative theorizing requires the idea that models are mathematical structures representing the systems under study. One does not simply embed experiments in some model of the theory, but gradually goes from qualitative finitistic structures to more complex ones. This fact helps us also in getting a clear picture in the hotly debated issue of scientific representation. To attribute to a model of the theory a direct role in accounting for the results of experiments may also lead us to forget the fact that representation is a relation holding (if it does hold at all) between target systems and data structures. It is not the theory which represents directly the behavior of the entities, but rather the data structures which reflect the (highly idealized) behavior of the entities in particular experiments. Of course, we do not have space here to enter into the details and intricacies of the debate concerning scientific representation, but we just point to that issue to make clear how foundational rigor may help in such kind of discussion.

So, it seems to us that CCA is a useful account of the semantic view. Its main force, however, is still not explored. Higher order model theory may provide some helpful insights into the structure of scientific theories. It is not possible for us, in the scope of this work, to develop further into those topics, but we agree with the general tone of Halvorson (2012): philosophers of science have the mathematical tools to study structures and models, we have only ourselves to blame if we turn our backs to the work. We now present an alternative account of the semantic approach, the one favored by Suppes when dealing with axiomatization.

5 The Suppesian approach

The Suppesian approach, as we mentioned before, differs from CCA by not having to employ a formal language to talk about the structures we wish to axiomatize: we do it directly in the set theory in which we built the structures themselves. Of course, most of the times this set theory is ZFC. In this case, despite the fact that we still want to characterize a class of structures as "models" of a theory, this time we cannot take those structures as models in the Tarskian sense, or in Thomson-Jones' truth making sense: there is simply no set of sentences to be made true by the structures according to an appropriate interpretation. Instead of that, the postulates are written in the language of set theory increased by some constants to allow us to talk about the relevant structure.

As Suppes has emphasized, it is 'more economic' to work as the present approach suggests, that is, presupposing a set theory (and its underlying logic), say ZFC, and not going to metamathematics (to the syntactical approach). Thus, it seems to be more interesting to *assume* a mathematics, say first-order ZFC set theory (but, as we have said, we could use other theories as well) and work *within* it. Suppes, as it is well known, works in an informal set theory, so it is not necessary to explicit neither the logic nor the set theory; in his approach, the things we need can be rescued by adding the required axioms to the set theory enrolled. Then, he says that "[a]s far as I can see, most problems of central importance to the philosophy of science can be discussed in full detail by accepting something like a standard formulation of set theory, without questioning the foundations of mathematic" (Suppes (2002), p.1). But *most* does not mean *all*, and there are interesting questions to be dealt with if we pay attention to the commitments assumed when we adopt a certain metamathematical basis, as we hope our discussion will make clear below.

To make the distinction clearer, let us once again present the case of group theory. As mathematicians usually characterize it, a group is an ordered pair $\langle G, \circ \rangle$, and axioms (G1) and (G2) above hold (see definition 4).

Now, before we proceed and present the usual set theoretical predicate for groups, let us recall how mathematicians prove, for instance, that $\langle \mathbb{R}, + \rangle$ is a group. That is very simple; once the set of real numbers is constructed inside set theory (with real numbers taken as Dedekind cuts, for instance), and addition is properly defined, one proves that the operation $+$ has properties G1 and G2. That is, one establishes the following theorems of ZFC:

- ZFC $\vdash \forall x, y, z \in \mathbb{R}, (((x + y) + z) = (x + (y + z)))$;

- ZFC $\vdash \forall x \in \mathbb{R}(x + 0 = 0 + x = x) \land \forall x \in \mathbb{R}, \exists y \in \mathbb{R}(x + y = y + x = 0))$.

Notice that in this case there is no need for an attribution of set theoretical entities to linguistic entities; there is simply no separate formal language designed to talk about groups. Instead of that, the derivations above are performed directly in the set theory, in this case increased by the new constant symbols $\mathbb{R}$, $+$, and 0. Indeed, there is no "satisfaction relation" holding between elements of the structure and formulas of a specific language. All the magic is performed inside ZFC proper. Of course, that also means that there are no specific model theoretic tools to deal with those structures either. What holds for groups holds also for more complicated theories, such as classical and quantum mechanics, so as for any theory whose models are set theoretical structures.

A major difference from the CCA approach concerns the availability of all the mathematics required in order to do science. We do not need to axiomatize the Dedekind complete field in order to speak about the field of real numbers, for we have them available from the underlying set theory (once again, provided by some particular construction, such as Dedekind cuts or Cauchy sequences). The same goes for vector spaces, Hilbert spaces, Euclidean geometry and the whole of mathematics required to do most of our current scientific theories. That fact may prompt the idea that mathematical structures in the Suppesian approach are already interpreted structures satisfying some postulates. Consider the simplicity of Suppes' style classical particle mechanics (CPM). A system of particle mechanics is a 5-tuple[10]

$$\mathfrak{P} = \langle P, T, m, s, f, g \rangle$$

where P is a non-empty set (the 'particles'), T is an interval in the set of real numbers (say expressing an interval of time), m is a function from P to $\mathbb{R}^+$ so that if $p \in P$, then $m(p)$ is the mass of p, s is a function from $P \times T$ in $\mathbb{R}^3$, so that $s(p,t)$ is a vector expressing the position of the particle p at time t, f is a function with domain $P \times T \times I$, where I is a set of positive integers, so that $f(p,t,i)$ is a vector representing the forces acting on p at t, g is a vector representing the external forces acting on a particle at a given time t. All these concepts are subjected to kinematical and dynamical axioms below:

Definition 5.1. A Newtonian particle mechanics is a structure

$$\mathfrak{P} = \langle P, T, s, m, f, g \rangle$$

satisfying the following axioms:

Kinematical axioms

P1 P is a finite non-empty set;

P2 T is an interval of real numbers;

P3 for $p \in P$, s_p is twice differentiable in T;

Dynamical axioms

---

[10]Here we take a simplified version of this structure; see Suppes (1957), chap. 12. In his (2002), Suppes provides an alternative formulation.

P4 for $p \in P$, $m(p)$ is a positive real number;

P5 for $p, q \in P, t \in T, f(p,q,t) = -f(q,p,t)$;

P6 for $p, q \in P, t \in T, s(p,t) \times f(p,q,t) = -s(q,t) \times f(q,p,t)$;

P7 for $p \in P$ and $t \in T, m(p)D^2 s_p(t) = \sum_{q \in P} f(p,q,t) + g(p,t)$.

The fact that we intuitively call P a set of particles, m a mass function and p a function attributing position to particles at every instant of time, should not allow us to forget that this is just an intended interpretation; what we really have here are abstract sets. Thomson-Jones (2006) calls this kind of presentation *description fitter*, in opposition to the previous one (CCA) which he calls, as we mentioned, *truth maker* models. The structures modeling the kinematical and dynamical axioms are not making a set of sentences *true*, but they are just fitting the description provided by the axioms. This is in close connection with our claim that in the Suppesian style axiomatization there is no formal language that gets interpreted in the structures in question, but rather the whole demonstration that the structure fits the description provided by the axioms, something that is performed inside the framework of set theory proper, using the deductive resources of this theory along with the particular features of the structure under scrutiny.

We may make this point even clearer by recalling an objection by Thomson-Jones against use of models in the Tarskian truth-maker sense. Recall from the previous section that the objection ran roughly as follows: since in model theory we always operate inside some set theory, the structures used as models are set theoretical entities, that is, they are limited to the sets the set theoretical apparatus is able to provide.[11] The objection advanced that the mentioned fact is really a limitation to the semantic approach, since one should not be restricted to the resources of some specific set theory. Now, as we are developing it, this objection could be made also against the Suppesian approach, since the structures taken into account here are also set theoretical entities. The mere fact that structures are being taken as description fitters, rather than truth makers, does not free us from building them inside some framework, a framework which must be strong enough to provide the mathematics required to develop scientific theories. So, the semantic approach will have to live with the fact that there is an underlying mathematical apparatus which must be recognized (and this is a central contention of this paper). Rather than being a limitation, as we

---

[11] Once more, some care must be taken when the theory is set theory itself.

mentioned in our answer to this objection in the previous paragraph, this fact may sometimes help us getting closer to scientific practice, as in the case of the finitistic structures required for a rigorous account of the data structures.

So, we should not forget that the intended interpretation of the structures, such as that one for classical particle mechanics, is really inside a set theory, which on its turn may be seen from a completely formal point of view as a first-order theory. Everything done in the set theoretical predicate for mechanics could in principle be rewritten only with $\in$. To get a clear idea of the framework we are working in is a necessary condition for foundational studies. Otherwise, how should we know what kind of tools we do have at our disposal? We shall soon exemplify how the apparatus we work in influences the discussion in foundations of science too.

The point then is that understanding structures as descriptions fitters does not exempt them from being set theoretical entities (nor entities in some mathematical framework). As we mentioned, this should not be seen as a limitation, even more because it is widely recognized that set theory is a very powerful framework, probably exceeding the needs of actual science (except, of course, the requirements of set theory itself, when considered as a branch of mathematics). The fact that set theory is the standard framework for the Suppesian approach is responsible for its great power (for further discussion see also Krause, Arenhart and Moraes (2011)).

Indeed, this enormous strength may be exemplified by a trivial example: as is well-known, we may easily present the Suppes predicate for well-orders as usually done by mathematicians. A well-order is a pair $\langle D, R \rangle$ such that R is a partial order over the non-empty set D and every non-empty subset of D has an R-least element. As an example of a well-ordered structure, take $\langle \omega, \in \rangle$. So, the class defined by the set theoretical predicate for well-orderings will have that structure as an element. Could we leave that structure out? Yes, it is really simple: just state another axiom requiring that the structures satisfying the set theoretical predicate should be well-orderings distinct from $\langle \omega, \in \rangle$. Of course, more interesting cases may be conceived, but this one already shows some of the resources that the Suppesian approach affords thanks to the use of the language of set theory.

The great expressive power of this language led Suppe to a little exaggerated praise of the Suppesian approach (*loc. cit.*). According to Suppe's interpretation of Suppes' approach to scientific theories, instead of presenting syntactic axioms we just select the intended models of the theory and work only with them. This avoids the objection presented in the previous section against the syntactic approach according to which a formalization in a formal first-order language is always prey to non-standard models (but recall our remarks on the scope of applicability of Löwenheim-Skolem and other standard model theory theorems). That is, according to this view, we escape non-standard models by selecting precisely the intended ones:

> In practice we *do* pick just intended models (else there would not be a problem of unintended models), and so we can proceed directly to the specification of these intended models without recourse to syntactic axiomatization.

However, one has to deal with unintended models in the Suppesian approach too. As one may easily notice, most models of a set theoretical predicate in the present approach will be mathematical models, having no relation to empirical reality. It is not difficult to find models for the postulates of particle mechanics above as structures that can be built within ZFC. Although we do not present them here, it is easy to acknowledge that they are just mathematical structures without having necessarily any connection to a field of knowledge in reality. But other models may constitute (as we believe) a map of parts of empirical reality (this is discussed in Suppes (2002)). Of course this presentation makes use of mathematics (ZFC set theory if we wish, and classical logic). But they are not explicitly stated, becoming implicit. Furthermore, if one is to block those unintended purely mathematical models somehow (say, by imposing new axioms which ban the unintended models), then the great advantage of Suppes' approach against the syntactical one gets diminished, since we are once again having to worry ourselves with the tricks required to keep only the intended models (see once again Suppe *loc. cit*).

The models of CPM are structures like $\mathfrak{P}$ above, and are set-theoretical constructs, structures in ZFC. The models, then, within this schema, are set-theoretical structures and what we can infer from them will depend on the underlying logic and on the axioms of set theory. It is not simply the case that we may pick whichever models we want. Let us just mention another example to illustrate this fact. In non-relativistic quantum mechanics, position and momentum operators are quite relevant. In the Hilbert space $L^2(\mathbb{R})$ of the equivalence classes of square integrable functions, they are unbounded operators. Just to recall, if A is a linear operator, then A is unbounded if for any $M > 0$ there exists a vector α such that $\|A(\alpha)\| > M\|\alpha\|$. Otherwise, A is bounded. But R. Solovay proved that if ZF (ZFC with the axiom of choice dropped) is consistent, and if DC stands for the axiom of dependent choices, then it follows that ZF (without the axiom of choice) plus DC has a model in which each subset of the real numbers is Lebesgue measurable.[12]

Let us call 'Solovay's axiom' (AS) the statement that "Any subset of $\mathbb{R}$ is Lebesgue measurable". Then, in the theory ZF+DC+AS (termed 'Solovay's set theory'), it can be proven that any linear operator is bounded — see [18]. Thus, if we use Solovay's theory instead of standard ZFC to build quantum structures, the mathematics of quantum mechanics would be different from the standard one. In particular, we are not sure how to deal with position and momentum operators.

Another example also involving Solovay is the following one. One of the fundamental theorems in functional analysis is Gleason's theorem (it does not matter for our purposes here its exact formulation), which is relevant in quantum mechanics. The theorem implies the existence of certain probability measures in separable Hilbert spaces. Solovay obtained a generalization of the theorem also for non-separable spaces, but it was necessary to assume the existence of a *gigantic orthonormal basis* whose cardinal is a measurable cardinal (see Chernoff (2009)). But the existence of such cardinals cannot be proven in ZFC set theory (supposed consistent). Thus, in order to get the generalization, we need to extend ZFC. That is, as far as a physical theory demands mathematics, perhaps new mathematical frameworks will be required (but this is not a novelty, for it has ever been this way!)

These examples show that for certain considerations, it is extremely important to consider the mathematical framework we are working in with great care. What counts as allowable models and the mathematics available in the construction of the relevant set theoretical predicate may vary according to the resources of the underlying set theory. Furthermore, one may come to the conclusion that some particular set theory is more convenient for some purposes than another. From the philosophical point of view, this is particularly true for expressing metaphysical views.

When other examples of such situations are taken into account, it may be the case that we arrive at a conclusion that ZFC seems no more adequate to be the ground of our science (or of some particular scientific theory). For instance, this would be the case if we find true contradictions in the world, as the *dialetheists* believe. Another example would be to strongly belief Birkhoff and von Neumann, who claimed

---

[12]DC is not sufficient to prove that there are nonmeasurable subsets of the real numbers.

that quantum mechanics would demand a non standard ('quantum') logic. In this particular setting, let us recall that the realm of *quantum logic* was erected from their seminal work, but although it has profound connections with quantum physics, quantum logics progressed as the algebraic study of certain lattices, without direct appeal to the foundations of quantum theory proper (for instance, see Dalla Chiara, Giuntini and Greechie (2004)). As is well known, some attempts to develop quantum mechanics from a non-classical logic were advanced, for instance, by Reichenbach (1998). Another specific case is that which shows that the standard quantum formalism is compatible with two alternative metaphysical packages concerning the nature of quantum objects; the first one takes them as *individuals* as in the classical sense, as entities that can be individuated and considered on a par with their "classical" cousins. Quantum objects of this sort can be treated inside ZFC, for sets are collections of such entities by definition. But there is another metaphysical package that comes from the forerunners of quantum theory, a view that sees them as *non-individuals*, as entities for which the standard notion of identity would not apply. These non-individuals can also be treated inside ZFC (which confirms the strength of this language), but at the expenses of losing the metaphysical assumption about their non-individuality. In this case, it has been said, a different mathematical framework would be suitable, namely, *quasi-set theory* (see French and Krause (2006) for a full discussion on this topic; see also Domenech, Holik and Krause (2008), Domenech, Holik, Kniznik and Krause (2010)).

In situations such as those above, we need to acknowledge that there are alternatives suggesting the use of deviant logics and mathematics in the realm of science.

6 Semantic view, foundations, and real science

As we have been discussing, there are two alternative ways to approach the semantic view. Both have their attractions and their defects. We believe that we have made clear that, according to our view on both types of semantic approach, the use of language is not an inessential part of the semantic view; in both approaches one should recognize that a class of models does not come by itself, but it must be characterized by some set of axioms; and these axioms have an impact on the nature of the class of models being defined.

Some may believe that, by allowing language to play such a fundamental role, we have betrayed the spirit of the semantic approach. However, we believe that only by putting the cards explicitly on the table we may have a clear picture of what is going on when relevant issues in philosophy of science are discussed. Indeed, recall that a view about the nature of theories is deeply relevant for the successful consecution of most philosophical purposes concerning science nowadays (to mention but a few: realism/anti-realism debate, structural realism and the nature of structure, issues concerning scientific representation, and so on); with so much at stake, it is only natural to try to have a most rigorous account of scientific theories. We believe our exposition has made it clear that there are many possibilities to be investigated, but the linguistic aspect must not be ignored. Now that some skepticism is being raised concerning the success of the semantic approach as the most adequate rendering of scientific theories, we must provide rigorous foundations for hand-waving slogans and verify the tenability of the view (for criticism of the semantic view, see Thomson-Jones (2006), Halvorson (2012)).

The first difficulty for those attempts (at least as we have proposed them) comes from a recent trend in de-emphasizing formal methods in philosophy of science. In some parts of the paper we have been concerned with criticisms as to the relation of some particular kinds of axiomatization with actual science. We have tried to diminish the weight of such criticism by making it clear how attention to axiomatization may be helpful in the study of actual science and may enlighten some aspects of it, in particular by pointing (with Suppes (1967), (2011)) that the use of constructive mathematics may be helpful to reflect the finitistic nature of the data structures obtained in scientific practice. That point is relevant in this situation because it has become part of the philosophical canon to dismiss a particular proposal by arguing that it does not reflect real science, mainly because the proposal in question amounts to what may be regarded as philosophically irrelevant "overformalization". Really, that position seems to be historically explained by recalling that the Logical Positivists' taste for logical analysis and the whole conceptual machinery required by their anti-metaphysical prejudices has created a general aversion to formalization and some forms of logical studies (see also the concluding remarks in Contessa (2006) for doubts on the usefulness of formal apparatus on philosophy of science; curiously, one of the criticisms Contessa advances to the particular approach he is examining in that paper is that it is not rigorous enough!).

That historical point notwithstanding, we would like also to argue that there is much in formalization that, despite the appearance of not being related to actual scientific practice, may be allowed by a scientifically respectful philosophy of science. One of those points, as we have emphasized above, comes from the use of diverse logical tools to understand the structure of scientific theories. Really, as philosophers of science we should be allowed to apply freely the techniques of logic and formal sciences in general for the investigation of topics of philosophical interest (even when that does not reflect "actual scientific practice"). Distinct models of ZFC, set theories employing weakened or restricted forms of the axiom of choice or even non-standard set theories should be investigated as frameworks for the semantic approach. That involves a great amount of formal work, but it should not be dismissed as being far away from any kind of scientific practice. Indeed, such formal investigation and detailed scrutinizing is what foundational work amounts to after all. Some idealization in the philosophical analysis of science is not a sin against scientifically engaged philosophy.

More than looking for the right formal apparatus to reflect some aspect of actual scientific practice, we must be aware that some topics of great philosophical interest involve taking a greater than usual care with the underlying framework. One of such cases concerns quantum mechanics. There are diverse aspects of the theory which require a discussion of the mathematical basis of the theory. To mention a few of them, we should point to (i) the attempts of capturing the Bohrian complementarity relation by some three valued logic, (ii) the attempts of modeling the non-distributive character of some logical operators by orthomodular lattices, (iii) the attempts of getting a clear picture as to the nature of individuality and non-individuality of quantum entities (see French and Krause (2006), Domenech, Holik and Krause (2008)). So, a look for the right tools may be of fundamental importance, and formalization in philosophy of science should not be feared.

For a second point of difficulty concerning the consecution of the project of endowing the semantic approach with rigorous foundation, we mention some consequences of the sloganeering that has characterized the approach most of the times until now. Most attempts to apply the semantic approach in philosophical projects have tended to *identify* a theory with a (supposed) linguistically-free class of models. So, theories were seen as classes of models properly. This is also claimed to be in accordance with scientific practice (recall Suppe *loc. cit.*). This trend have led to two kinds of related objections to the semantic approach. The first one concerns obvious difficulties with the identity of scientific theories, a problem that resemble the difficulties that plagued the syntactic approach. Let us call this the *Identity Problem*. The second main difficulty concerns the central role attributed to theory in scientific practice and in philosophical investigations of science. Let us call this the *Theory Monopoly Problem*.

As we deal with the Identity Problem in section 4, the difficulty appears that distinct kinds of structures may be models of what we would (by independent intuitive criteria) call *the same* theory. So, the approach, by making the strict identification of a theory with a class of models seems to take as distinct theories those classes of structures which should present the same theory (see the example of group theory in section 4). A more liberal view of the role of language may help us in avoiding such troubles (see the related concerns in Muller (2011) and Halvorson (2012)).

Furthermore, in our discussion throughout the paper we have not held that a theory *is to be identified with a class of structures*, but rather that a theory is *presented* by a class of structures. That is, one may investigate some relevant aspects of the theory by dealing with the class of models. Notice, this is much more akin to scientific practice, where scientists know, at least most of the times, when they are dealing with the same theory. Our point is that there is an intuitive body of knowledge which may be rendered rigorous by diverse classes of models, all of them presenting the same theory. So, the strong thesis attributed to the semantic view, the one claiming that *a theory is a class of models* should be weakened to contemplate distinct presentations of what would intuitively be regarded as the same theory. Obviously, as we have suggested before, this conditions for the identity of a theory could be made rigorous by the adoption of the condition of equivalence between vocabularies, where equivalence involves definability of the terms of one vocabulary in terms of the other, and vice-versa.

For another way of putting that issue, one could distinguish two related senses of 'identify'. First of all, we do not identify a theory *with* a class of models, as it was made clear in the previous paragraphs; this is not required by the semantic approach. What one may reasonably do, however, is to identify a theory *by* a class o models. In this sense, the required class of models may be employed to identify the theory one is dealing with, but it is not the theory itself. As we have been discussing, this distinction avoids the problems of distinct classes of models being employed to present the same theory, where the term 'theory' is now employed in an informal sense[13].

As for the Monopoly of Theory Problem, we must be aware that scientific theorizing and working with the development of theories is only part of the real scientific enterprize. There is much more to science than that, of course. Scientists conduce experiments, rely on intuitive thinking, lucky guesses and hypothesis, on tacit knowledge and a lot of other practices that do not always involve theorizing. Most of those aspects of science, in particular a lot of social aspects, are not involved in the

---

[13] We would like to thank to an anonymous referee for pointing this to us.

schema provided by the semantic view. One should recognize that the semantic view concerns the logical aspect of the structure of theories, but leave most of the other aspects of science untouched. Otherwise, we shall fall prey to the criticism that the semantic view does not provide for a complete account of science. However, as we have been mentioning, only some aspects are to be touched on by the semantic view, and we must take a humble stance toward that fact, recognizing that the scientific practice involves other aspects as well. This by no means diminishes the importance of the view, but may be used to recall that it was designed precisely as a view about *theories*.

BIBLIOGRAPHY

[1] Arenhart, J. B.; Moraes, F. T. F., Structures, languages and models: a unifying approach. *Logique et Analyse*, 56(221), pp. 67-84, 2013.
[2] Balzer, W., Moulines, C.U. and J.D. Sneed, *An Architectonic for Science. The Structuralist Program*, Dordrecht: Reidel, 1987.
[3] Chakravartty, A., The semantic or model-theoretic view of theories and scientific realism. *Synthese*, 127, pp. 325-345, 2001.
[4] Carnap. R., *Introduction to Logic and its Applications*. Dover, 1958.
[5] Chernoff, P.R., Andy Gleason and quantum mechanics, *Notices of the American Mathematical Society*, November, pp.1253-1259, 2009.
[6] Church, A., *Introduction to Mathematical Logic*. Vol. 1, Princeton: Princeton Un. Press, 1956.
[7] Contessa, G., Scientific models, partial structures and the new received view of theories. *Studies in History and Philosophy of Science*, 37, pp. 370-377, 2006.
[8] da Costa, N. C. A.; Chuaqui, R., On Suppes' Set Theoretical Predicates. *Erkenntnis*, 29, pp. 95-112, 1988.
[9] da Costa, N. C. A. and Rodrigues, A. A. M., Definability and Invariance. *Studia Logica*, 82, pp. 1-30, 2007.
[10] Dalla Chiara, M.L., Giuntini, R. & Greechie, R. *Reasoning in Quantum Theory: Sharp and Unsharp Quantum Logics*, Kluwer Ac. Press (Trends in Logic, vol.22), 2004.
[11] Domenech, G., Holik, F. and Krause, D. Q-spaces and the foundations of quantum mechanics, *Foundations of Physics* 38 (11), 969-994, 2008.
[12] Domenech, G., Holik, F., Kniznik, L, and Krause, D., No Labeling Quantum Mechanics of Indiscernible Particles, *International J. Theoretical Physics* DOI 10.1007/s10773-009-0220-x, 2010.
[13] French, S., and Krause, D., *Identity in Physics. A historical, philosophical and formal analysis*. Oxford: Oxford University Press, 2006.
[14] Halvorson, H., What scientific theories could not be. *Philosophy of science*, 79(2), pp. 183-206, 2012.
[15] Hinst, P., A Rigorous Set Theoretical Foundation of the Structuralist Approach. In Balzer, W. and C.U. Moulines (eds.), *Structuralist Theory of Science. Focal Issues, New Results*, Berlin: de Gruyter, pp. 233-263, 1996.
[16] Krause, D.; Arenhart, J. R. B.; Moraes, F. T. F., Axiomatization and models of scientific theories. *Foundations of Science*, 16, pp. 363-382, 2011.
[17] Krause, D.; Arenhart, J. R. B., A discussion on quantum non-individuality. *Journal of applied non-classical logics*, 22, pp. 105-124, 2012.
[18] Maitland Wright, J. D., All operators on a Hilbert space are bounded, *Bulletin of the Americal Mathematical Society* 79 (6), 1247, 1973.
[19] Mormann, T., Categorial Struturalism, in W. Balzer and C.U. Moulines (eds.), *Structuralist Theory of Science - Focal Issues, New Results*, Berlin/New York: Walter de Gruyter, pp. 265-286, 1996.
[20] Muller, F. A., Reflections on the revolution at Stanford. *Synthese*, 183(1), pp. 87-114, 2011.
[21] Reichenbach, H., *Philosophic Foundations of Quantum Mechanics*. New York, 1998.
[22] Stegmüller, W., The Structuralist View of Theories, Berlin: Springer, 1979.
[23] Suppe, F., *The Structure of Scientific Theories*. 2nd. ed., Urbana and Chicago: U, Illinois Press, 1977.
[24] Suppe, F., Understanding scientific theories: an assesment of developments, 1969-1998. *Philosophy of science*, 67 Supplement, pp. S102-S115, 2000.

[25] Suppes, P., *Introduction to Logic*. Toronto: van Nostrand, 1957.
[26] Suppes, P., A Comparison of the Meaning and Uses of Models in Mathematics and the Natural Sciences, *Synthese* **12**, pp. 287-301, 1960.
[27] Suppes, P., Models of data. In: E. Nagel, P. Suppes and A. Tarski (eds.), *Logic, Methodology and Philosophy of Science: Proceedings of the 1960 International Congress*, Stanford: Stanford Un. Press, 1962, pp. 252-261.
[28] Suppes, P., What is a scientific theory? In: Sidney Morgenbesser (ed.), *Philosophy of Science Today*, New York: Basic Books, 1967, pp. 55-67.
[29] Suppes, P., *Representation and Invariance of Scientific Structures*. Stanford: CSLI Pu., 2002.
[30] Suppes, P., Future develpment of scientific structures closer to experiments: response to F. A. Muller. *Synthese*, **183**(1), pp. 115-126, 2011.
[31] Thomson-Jones, M., Models and the semantic view. *Philosophy of Science*, **73**, pp. 524-535, 2006.

Between Weasels and Hybrids: What Does the Applicability of Mathematics Tell us about Ontology?

STEVEN FRENCH

ABSTRACT. What are the implications of the significant and ubiquitous role of mathematics in physical explanations? I shall consider two answers to this question. The first takes this role to imply the existence of mathematical objects; the second takes it to imply that certain physical quantities have a 'hybrid nature'. I shall close off one line of defence with regard to the first, by arguing that understood in its appropriate metaphysical context, this line loses its force. A similar focus on the relevant metaphysics allows us to block the second. The conclusion drawn is that attention needs to be paid to these metaphysical considerations if the debate over the role of mathematics is to move forward.

1 Introduction

What are the implications of the role of mathematics in scientific explanations? One answer is that it implies the existence of mathematical objects. According to the 'Enhanced Indispensability Argument' (EIA) (see Baker 2009; Daly and Langford 2009; Saatsi 2011), we ought rationally to believe in the existence of those mathematical entities that play an indispensable role in physical explanations. Two of the most well-known examples that have been given of this apparently indispensable role are honeycomb shape and cicada life cycles. Let me very briefly consider each in turn.

2 Examples

i) *Honeycombs*

Here the relevant explanandum is the hexagonal shape of bee honeycomb cells, and the relevant mathematical explanans states that the optimal shape for tiling the Euclidean plane (in the sense of the minimum total side length for arbitrarily large areas) is a hexagon (Hales 1999). However, due to concerns over appropriately drawing the distinction between physical and mathematical geometry, the focus has shifted to:

ii) *Cicadas*

Here the explanandum is the prime periodic life cycles of cicadas in the U.S. (13 and 17 years depending on the species) and the explanans includes the conjunction of the number-theoretic fact that prime periods minimize the relevant intersection and the biological claim that a life cycle period that minimizes intersection (with other periods) is evolutionarily advantageous. Now, in this last case, one might well wonder what it is one is expected to rationally believe here: is it simply the existence of the numbers 13 and 17, or of all prime numbers or all numbers? Consider for example the sequence of explanations involved in showing that the negatively charged particles (now identified as electrons) that explained the phenomena revealed in certain experiments on 'cathode rays' were also produced by heated, illuminated and radioactive materials. In such cases the explanans typically involves the existence of a kind of thing, i.e. electrons, the positing of which is extended to cover other explanandums. In the cicadas case, although we can envisage the initial positing being likewise extended to numerous other cases, it is just not clear what we are being invited to accept. By comparison with the electron case, we might take it to be the kinds '13' and '17' but with-holding belief in other kinds, although appropriate in the electron case, does not seem so here: once we have established the existence of the number 13, it seems we can quickly infer that of 12, 14, and the rest of the reals (since, of course, one of the fundamental properties that numbers possess is being related by succession). What the explanation thus gives us grounds to believe in the mathematical case is a certain number structure. This might suggest that the advocate of the EIA should be a structural realist when it comes to mathematics. However, the EIA itself can be resisted by denying that the relevant mathematics actually plays any explanatory role at all.

3   Indispensability and Indexing

Thus, perhaps the well-known response to the argument insists that mathematics has only an 'indexing' (Melia 2000) or representational role (Saatsi 2011), rather than an explanatory one. Melia writes,

> '[Although we may express] the fact that a is 7/11 metres from b by using a three place predicate relating a and b to the number 7/11, nobody thinks that this fact holds in virtue of some three place relation connecting a, b and the number 7/11. Rather, the various numbers are used merely to index different distance relations.' (Melia 2000, p. 473)

Similarly, Saatsi argues,

'Mathematics can help us learn about the world by virtue of playing a representational (and, derivatively, inferential and justificatory) role in science. The fact that mathematics can play such a role is something that calls for an explanation. But this issue, and any argument for Platonism that turns on this issue, is independent from EIA. The latter concerns not the representational capacity of mathematics in general, but more specifically the capacity and role that mathematics per se has in explaining and yielding understanding about the concrete world.' (Saatsi 2011, p. 145).

Saatsi in particular includes details of how an explanation can be given of the cicada example that appears to dispense with the relevant mathematics by, for example, using physical rods of a certain length to represent cicada lifetimes (Saatsi op. cit.).

Of course, it may be that such non-mathematical representations are not (straightforwardly) available in all cases. At least at first thought it might seem difficult to envisage how Saatsi's manoeuvre will work for more complex examples, such as we find in physics, and as indicated below. In such cases the mathematical and (for want of a better phrase) 'the physical' are so entwined that its at the very least unclear how the two might be separated in such a way that the mathematics can be dispensed with for explanatory purposes, as it can in the cicada example. Indeed, some have argued that in at least some highly significant cases in physics (such as those that involve the behaviour of systems near phase transitions; see Batterman 2010) no such separation can be achieved and the mathematics must be accorded an explanatory role (but see Bueno and French 2011).

4 Representation and Explanation

Let me note two points at this juncture. The first is that this sense of entwining may be encouraged by the use of mathematics itself as a representational device by philosophers of science, following Suppes' famous admonition that they should use mathematics not meta-mathematics in this regard. This forms one of the cornerstones of the so-called 'semantic' or model-theoretic approach to theories founded by Suppes (among others) and has led some commentators to conclude that this approach *identifies* theories with mathematical, or more precisely, set-theoretical entities and to criticise it on those grounds.

Certainly I would agree that the semantic approach (suitable modified) offers a useful way of *representing* the relationship between scientific theories and the relevant mathematics and hence provides an appropriate framework in terms of which various issues regarding the applicability of mathematics can be analysed and discussed (see for example Bueno 1999; Bueno, French and Ladyman 2002; Bueno and French forthcoming). In particular, representing, at the (meta-) level of the philosophy of science, the representational capacity of mathematics within science by means of partial homomorphisms couched in terms of 'partial structures' (da Costa and French 2003) allows us to capture the sense in which mathematics provides 'surplus structure' in this context, as well as the associated heuristic fertility and the applicability of mathematics more generally (Bueno and French forthcoming). In particular, this approach allows for the carrying over of relevant structural features from the mathematical level to that of the physical theory, as in the further example of the explanation of superfluidity in terms of Bose-Einstein statistics and, hence, via the understanding of the latter in terms of symmetric state functions, group theory (Bueno, French and Ladyman 2002). But of course, even if one accepts this framework that does not imply that one should accept that the relevant mathematics is explanatorily indispensable in the way that would feed into the EIA and support the ontological conclusion.

Returning to the claim that the entwining of mathematics and (modern) physics supports the explanatory indispensability of the former, my second point is that in response one might insist that the onus is then on the advocate of such a view to articulate the sense in which mathematics can play such an explanatory role, in particular by situating it within one or other of the extant accounts of explanation in science, for example (see Bueno and French 2011). I shall return to this issue shortly.

5 Welcome to the Weasel

Furthermore, even in the face of this representational entwining, one can reject the implication of explanatory indispensability in various ways. Thus one might adopt Melia's 'weaseling' manoeuvre: we posit certain mathematical entities in order to construct the appropriate representation, only to subsequently deny the existence of these entities. The mathematics is then seen as '... the necessary scaffolding upon which the bridge must be built. But once the bridge has been built, the scaffolding can be removed.' (Melia op. cit.) Support for this manoeuvre can be sought by pointing to the role of idealisations in science: we may use the axioms of three-dimensional Euclidean geometry as the 'scaffolding' by which we define a two-dimensional non-Euclidean surface, for example (ibid.). Another example would be the initial deployment of rigid rods in the development of Special Relativity, only for this deployment to be rescinded once the theory had been constructed (indeed, the theory itself denies the physical possibility of rigid rods; see for example Shapere 1969).

Unfortunately things are not quite so straightforward as the proponent of 'weaseling' might like. In both the above cases we can reformulate the theory in such a way as to eliminate the idealisation: by describing the non-Euclidean surface intrinsically, or rewriting Special Relativity in terms of space-time structure. It is not clear that we can do something similar when it comes to the mathematics. Cicada lifetimes are one thing, lightcones quite another. Can we dismiss the relevant mathematics as mere scaffolding in the latter case? Again, and as we shall see, in certain cases, it can be claimed that the mathematics cannot be separated out in this manner since, it is argued, it constitutes an aspect of the very nature of the property concerned.

6 Explanatory frameworks

Now, although I shall indicate how this argument might be resisted, let me concede for now that the mathematics may be more than mere heuristic scaffolding. Nevertheless, that does not imply that it plays an explanatory role. At the very least, as I've already indicated, it needs to be shown how that role meshes with some account of explanation in science.

In this context, the defender of EIA can easily (and not surprisingly) expand the remit of the Deductive-Nomological approach to explanation to include mathematical laws and principles (Baker 2005). However this still does not explicate how such laws and principles can themselves have explanatory power (see Daly and Langford op. cit.). The D-N approach also suffers from well-known deficiencies, but appealing to alternative causal accounts of explanation to rule out an explanatory role for mathematics is obviously a question begging move.

Strevens' 'kairetic' approach (2008) may provide the defender of EIA with an entry point here. This places the focus on difference-making, in virtue of which only certain causal influences can be deemed as relevant to the occurrence of a phenomenon. Using an optimizing procedure to identify these difference makers, Strevens is able to show how irrelevant influences can be eliminated and a stand-alone explanation obtained in which one state of affairs bears causal-explanatory relevance to another.

Mathematics then qualifies as an 'explanatory tool' because it is through grasping mathematical dependences and independences that we are able to grasp *causal* dependences and independences (ibid, 331; see also Pincock forthcoming). Now, an immediate objection here is that functioning as an explanatory tool is not quite the same as playing the sort of explanatory role that supports the kinds of existence claims associated with the Indispensability Argument. Indeed, allowing us to grasp causal dependences is precisely one of the things that representational devices do. Furthermore, when it comes to the core claim of Strevens' account, it might be insisted that mathematics makes no difference when it comes to physical states of affairs and hence cannot bear the kind of explanatory relevance that Strevens has in mind.[1]

However, Baker has rejected as fatally flawed the arguments that have been given in support of the claim that mathematics makes no difference and argued that indispensability considerations are in fact crucial to the evaluation of this claim (Baker 2003). In particular the intuitions that typically lie behind it rely on an implicit analogy between abstract objects and remote, concrete ones. The latter, when posited by a theory, get dismissed as dispensable, precisely because their failure to make a difference is indicative of their lack of causal role. A similar line of reasoning runs the risk of question-begging when it comes to mathematical objects and Baker concludes that if the mathematical objects concerned are taken to be indispensable, then there is no determinate answer to the question whether their existence makes a difference.

Moreover, the advocate of EIA might see a way out of this impasse in Strevens' comment that the central difference-making criterion '... takes as its raw material any dependence relation of the "making it so" variety, including *but not limited to causal influence* (Strevens op. cit., 179; my emphasis). The idea is that once we have established the relevant dependence relation between some state of affairs and some set of 'entities', the criterion will tell us what facts regarding those entities underpin the relation's 'making it so' (ibid.). If this is not limited to causal influences, then the defender of EIA could adopt this approach and insist that there is a sense in which the mathematics 'makes it so'.

7 Another Example: White Dwarf Collapse

A useful and illustrative example, given in response to appeals by the critics of EIA to the standard causal account of explanation, is that of the explanation of the halting of the gravitational collapse of white dwarf stars by Pauli's Exclusion Principle (Colyvan forthcoming). Here the explanandum is the cessation of the collapse of the star under gravity and the explanans is the Exclusion Principle which determines how many electrons can occupy the relevant energy state. The core of the explanation usually given is that the gravitational attraction that is com-

---

[1] I am grateful to Juha Saatsi for raising this point.

pressing the stellar material is balanced by what is sometimes called the 'Pauli pressure', or 'degeneracy pressure' created by the occupancy of the energy states as dictated by the Exclusion Principle. In essence, the collapse stops because no more fermions can be packed into the available states. Now, insofar as the Principle cannot be regarded as a causal law, it can be claimed that this represents an example of an acausal explanation of the behaviour of physical systems that opens the door to the explanatory role of mathematics. Indeed, the defender of EIA can argue that, within the framework of the kairetic account, it is the Exclusion Principle that 'makes it so' with regard to the halting of stellar collapse and that, given an appropriate construal of this principle, this exemplifies the relevant indispensable explanatory role of mathematics.

The strength of such a claim obviously depends on how we understand the Exclusion Principle itself (see Massimi 2005). Here philosophy has not served us well in providing a useful analysis. Lewis, for example, talks of the Principle as representing 'negative information':

> 'The state-space of physical possibilities gave out. ... [I]nformation about the causal history of the stopping has been provided, but it was information of an unexpectedly negative sort. It was the information that the stopping had no causes at all, except for all the causes of the collapse which were a precondition of the stopping. Negative information is still information.' (Lewis 1986, 222-23)

Although this could be too easily dismissed as suggesting that the lack of causal information is still indicative of causal relevance (Colyvan op. cit.), the initial suggestion regarding restrictions on the state-space is at least on the right track, as we shall shortly see.

Strevens also considers this example and the role of the Principle within his kairetic approach:

> 'What relation holds between the law [PEP] and the arrest, then, in virtue of which the one explains the other? Let me give a partial answer: the relation is, like causal influence, some kind of metaphysical dependence relation. I no more have an account of this relation than I have an account of the influence relation, but I suggest that it is the sort of relation that we say "makes things happen".' (Strevens 2008, 178)

If the Pauli Exclusion Principle (understood acausally) can 'make things happen' via some kind of 'metaphysical dependence relation', then, the defender of EIA might ask, why can't mathematical objects? In particular, why should grasping mathematical dependences and independences be indicative of our grasp on causal dependences and independences only? Why couldn't this grasp reveal a dependence relation holding between the physical state of affairs and the relevant mathematical entities themselves?

8 Explanation via Symmetry

One way of responding to these questions is to indicate how we can accommodate the explanatory role of principles like Pauli's without introducing a vague and unexplicated metaphysical dependence relation. And this appears to be straightforward: the Exclusion Principle can be regarded as simply a manifestation of the appropriate symmetry requirement for particles obeying Fermi-Dirac statistics, namely that the relevant wave function be anti-symmetric; for those obeying Bose-Einstein statistics is must be symmetric. These alternatives correspond to two of the representations (the anti-symmetric and the symmetric respectively) of the permutation group and so Pauli's Principle drops out of the formalism associated with what has been called Permutation Invariance, that in turn can be understood as a fundamental symmetry principle of quantum mechanics (see French and Rickles 2003; French and Krause 2006).[2] Of course, how one understands the symmetry is crucial. Huggett draws on the parallel between permutations and covariant spatial transformations and constructs a framework in which quantum statistics (and hence Pauli's Principle) emerge as '... a natural result of the role symmetries play in nature.' (Huggett 1999, 346 for discussion see French and Rickles 2003). Thus he argues that it is implied by the conjunction of a further symmetry principle obeyed also be spacetime symmetries together with the formal structure of the permutation group. This further principle is what he calls "global Hamiltonian symmetry" which implies that the relevant symmetry operator commutes with the relevant Hamiltonian. With regard to the permutation group, of course, permutations of a subsystem are permutations of the whole system and this "global Hamiltonian symmetry" very straightforwardly implies Permutation Invariance, (Huggett, 1999b, pp. 344-5). Hence, Huggett concludes (ibid., p. 346), we should view permutation invariance as a particular consequence of global Hamiltonian symmetry given the group structure

---

[2] In non-relativistic quantum mechanics the symmetry is imposed on the theory; in algebraic quantum field theory it arises 'naturally' as a result of the imposition of a certain selection criterion on the set of representations of the permutation group; see Halvorson and Müger 2006.

of the permutations (of course, the issue remains as to the status of that group structure; again see French and Rickles, op. cit.). As a result the relevant Hilbert space can be thought of as divided up into sub-spaces, corresponding to the different group-theoretic representations and hence different statistics (Fermi-Dirac statistics obtain when the particles 'sit' in the symmetric sub-space, and Bose-Einstein statistics when they sit in the anti-symmetric sub-space). Given the symmetric nature of the appropriate Hamiltonian, once 'in' such a subspace, particles cannot get out, as it were. It is in this sense that Lewis was on the right track in focussing on the constraints on the relevant state space: from this perspective, the Exclusion Principle can be thought of as an expression of the 'limits' placed on the Hilbert space through the requirement that the relevant system be permutation invariant. Now, this does not completely close the door on the explanatory role of mathematics since Permutation Invariance (and symmetry principles in general) could be construed as purely mathematical and in this sense one could argue that mathematics plays such a role via imposing the above requirements. Thus what is needed is an understanding of such principles as representing physical aspects of the world. Fortunately there are various ways in which such an understanding might be motivated (Brading and Castellani 2003). Indeed, these principles can be tied to conservation laws in certain cases (see Brading and Brown 2003) and certain of them have been taken to underpin the classification of elementary particles in high energy physics.[3] Whether the principles are interpreted as imposing higher-level constraints on physical laws[4] and the kind structure of elementary particles, in the way that Permutation Invariance is typically interpreted as doing, or as regarded as properties of the laws or in some other way, they can be understood as physical, not mathematical. Furthermore, even though our understanding of the above kinds and physical quantities such as angular momentum and spin has been taken to derive in considerable part from their group-theoretic representation (Mirman 1995)[5] this is not sufficient to support the conclusion that these quantities are in some (non-representational) sense wholly mathematical[6] or that mathematics is playing the appro-

---

[3] As in the case of the so-called 'space-time' and internal symmetries, such as those associated with the Poincaré group and the $SU(3) \times SU(2) \times U(1)$ symmetry of the Standard Model respectively.

[4] Wigner — who did so much to articulate the role of such principles in physics — appears to have held both of these positions.

[5] Guay and Hepburn (2009) consider the extension of this formalism to groupoids and advocate a form of 'middle-ground' involving equivalence classes.

[6] Not least because different physical quantities may be described by isomorphic representations, as in the case of spin and isospin. One option would be to regard the

priate role in explanations involving these quantities. Certainly, as I've already emphasised, the onus is on the advocate of the EIA to show otherwise and in particular to articulate that role within an appropriate account of explanation.[7]

Returning then to the Exclusion Principle and stellar collapse, although we should avoid any tendency to think in terms of some kind of exclusion 'force' operating in this case (see Mullin and Blaylock 2003), and can accept that the explanation is acausal in some sense, it is not so in a way that immediately provides support to the defender of EIA. The role of symmetry principles in explanation deserves more discussion than we can give it here[8] but thinking of this role in terms of the imposition of constraints on the relevant dynamics, for example, enables us to understand such cases without either having to invoke 'negative information' or appealing to mathematics as explanatory. In the case of stellar collapse, the anti-symmetrisation of the relevant wave-function for the assembly *constrains* the distribution of electrons among the available states, such that they are effectively 'pushed' into occupying higher energy states, which leads to an apparent counter-gravitational pressure.

The 'metaphysical dependence' to which Strevens alludes can now be understood as underpinning these constraints, where it is the latter, and hence the associated symmetry, that 'makes things happen' (or not). How one understands both the constraints and the dependence will depend on one's ontology:[9] the structural realist, for example, will take the dependence to hold between the physical structure and the relevant putative entities, processes and regularities (French

---

quantities as not wholly mathematical but as hybrid in the sense that I shall explore and reject below in the context of the specific example of spin. I am grateful to a reader of an earlier version for pressing me on this point.

[7] Thus one might try to articulate that role within the 'unificationist' account of explanation which holds that explanation has to do with the 'bringing together' of previously disparate phenomena within some common framework (Friedman 1974; Kitcher 1989). Certainly mathematics can be seen to play a significant role within unificatory efforts in science but, again, the nature of this role requires articulation. One can argue, again, that it is essentially only representational, in that it can be the case that by casting theories in common mathematical form one can discern the commonalities between them that motivate and power attempts at unification. Given that, the onus is once again on the advocate of EIA to show that mathematics' role is more than this. Again I am grateful to one of the referees for urging me to consider this option.

[8] Dorato and Fellini (forthcoming) revive a form of 'structural explanation', in the context of which they suggest that that the relevant mathematics can play an explanatory role. However, it is clear that they do not mean this in the sense that defenders of EIA do. It may be that the role played by symmetries can be captured within their account.

[9] A useful overview of the metaphysics of dependence can be found in Correia 2008.

forthcominga). Of course, it isn't *necessary* to go down this route, as long as these constraints are understood as reflecting physical reality and an appropriate metaphysics is then given.[10] Dispositionalists who reject essentialism (which typically precludes the imposition of such constraints; see Bird 2007) and make room for symmetry as a feature of the world can also accommodate the kind of dependence involved here.

Chakravartty's 'semi-realism' is a case in point with regard to this last option: here the claim is defended that the kinds of structures we should be realist about are concrete, and conceived of in terms of relations holding between first-order properties of things (2007, p. 41). These first-order properties are, crucially, causal and one can epistemically delineate a sub-set of 'detection' properties; that is, properties that are 'causally linked to the regular behaviours of our detectors' (ibid., p. 47) and '... in whose existence one most reasonably believes on the basis of our causal contact with the world.' (ibid.). Furthermore, these properties should be understood in terms of dispositions for specific relations which comprise the concrete, physical structures. Indeed, according to the dispositional identity thesis (DIT) the identity of these causal properties is given by the dispositions they confer. In addition a notion of 'sociability' is introduced to account for the clustering of properties one observes. As applied to the kinds we obtain in physics (such as fermions and bosons), this can then be extended to cover symmetries and thus the constraints represented by the latter come to be understood in terms of this metaphysical notion of sociability with an associated dependence holding between the relevant dispositions.[11]

My aim here is not to compare and contrast these different accounts but merely to point out that they are available. The onus is then on the defender of EIA to show that not only has the problem just been pushed back a step but that no such physical dependence holds and that these constraints must be understood as mathematical. Given the availability of these alternatives, the prospects of such a demonstration

---

[10]This latter point is crucial of course, since some versions of structural realism take the relevant structures to be presented via the appropriate mathematics, such as group theory, which threatens to return mathematics to a fundamental explanatory role. What the structural realist has to do is articulate her position in such a way that this presentation at the level of physical theory can be understood as representational only, where the physical structure that is being represented is then articulated in appropriate metaphysical terms. Again I am grateful to a reader of a previous version for urging me to consider this point.

[11]Whether sociability is playing a robust explanatory role here or is merely piggybacking on the role played by the symmetry principles themselves is a matter for debate. However this is not my concern here (see French forthcoming b).

do not look good.

9 Hybridity

In response one might retreat to an alternative response to the entwining of science and mathematics which does not take such entwining to support the existence of mathematical entities as a distinct ontology but rather that certain physical objects and/or the associated quantities have a 'hybrid' mathematical-and-physical nature.

Early suggestions along these lines can be found in Heitler?s reflections on the role of 'profound mathematical concepts', such as group theory, in quantum mechanics which led him to conclude that the atom, for example, '... can hardly be thought of as something of a purely material nature' (Heitler 1963, 53). The idea seems to be that it is not just that we have to give up some kind of substance based ontology, or accept wave-particle duality in one or other of the well-known ways, but that the role of the afore-mentioned concepts implies that the very nature of the atom is in some sense partly mathematical. Given that the kinds of symmetries touched on above are represented mathematically via group theory one could invoke these examples as support for Heitler's view.

10 Yet Another Example: Spin

More recently, Morrison has drawn on similar considerations to argue that a particular physical property, namely spin, should be understood as 'hybrid' in that it possesses both mathematical and physical features and thereby 'bridges' the mathematical and physical domains (Morrison 2007).[12] It is widely acknowledged that spin cannot be conceived in classical terms but Morrison articulates the complex combination of both theoretical and experimental moves that afford us an appropriate grasp of it.

In the case of spin, the role of the relevant mathematics is particularly clear, since the property effectively 'drops out' of the Dirac equation, in the sense that,

---

[12] The two cases considered here — the Exclusion Principle and spin — can obviously be related via the spin-statistics theorem. Although there remains some doubt over what counts as an adequate proof of this theorem (see Sudarshan and Duck 2003), one could interpret it as grounding the relevant statistics in an understanding of spin, thus removing the need to appeal to symmetry as playing a fundamental explanatory role. Of course, this provides succour to the defender of EIA only if the weaker claim can be sustained. Furthermore, Berry and Robbins' 'geometric' proof turns the grounding relation in the other direction, insofar as on their account 'exchange involves hidden rotation' (Berry and Robbins 1997). This important approach to the theorem still awaits philosophical analysis.

'... the mathematical formalism of the Dirac equation and group theory require the existence of spin to guarantee conservation of angular momentum and to construct the generators of the rotation group.' (Morrison 2007, 546).

Although this provides a more complete theoretical treatment of spin than previous accounts, she regards it as 'essentially', but not wholly, mathematical. Such a view is reinforced by consideration of the group-theoretic analysis of spin (in the context of the classification of elementary particles), that helps to underpin our current understanding of this property (ibid., 552). Nevertheless, despite the problems with measuring single spins,[13] it can still be regarded as a measurable feature of the world in some indirect sense (ibid.). Thus, spin 'bridges' the gap between the mathematical and the physical and ... emerges as a peculiar hybrid notion possessing both physical and mathematical features.' (ibid., 547; also 552). And Morrison takes this as symptomatic in that, '... the nature of both theory and experiment in contemporary physics has largely stripped us of the resources for a sharp division between the mathematical and the physical.' (ibid., 555)

As a result, she insists, any attempt to give a realist interpretation of spin would be fundamentally misguided (ibid., 552-54). My concern here, however, is not primarily with this issue but with that of the grounds for regarding properties such as spin as hybrid in the first place. It certainly cannot have anything to do with the Galilean point about mathematics as the language in which physics is written, since that concerns only the representational role and would reduce the claim of hybridity to triviality.

11 Justifying Hybridity

One such set of grounds concerns the problems - discussed extensively by Morrison - with measuring spin and, in particular, the spin of, for example, single electrons. As she notes, these problems could be a result of some physical restriction on the very possibility of such observations, akin to that which lies behind quark confinement for example, or it could simply be a matter of technological difficulties (ibid., 552). However, the lack of any currently acceptable, theoretical motivation for supposing the former, together with recent experimental developments, suggests the latter. Indeed, as she herself notes, the use of magnetic force resonance microscopy had got the relevant sensitivity down to around 100 electron spins at the time she considered this issue,

---

[13]Problems that, as she says, enhanced its status as a truly non-classical feature of the world (ibid., 555).

but further work in this area has apparently demonstrated the ability to measure single spins (via the use of ultrasensitive cantilever based force detection sensors to measure the force resulting from a precessing spin in a magnetic field, together with an increase in the relevant theoretical understanding of, for example, spin relaxation processes; see Rugar et. al. 2004[14]).

Alternatively, and perhaps more productively, we might consider the purportedly 'essentially mathematical' nature of the property arising from Dirac's derivation. This kind of claim crops up at various places in the literature. Thus Steiner (who Morrison cites as providing an account of the relationship between mathematics and physics that might accommodate her conclusions; op. cit., 548, fn. 29), describes Dirac's discovery as 'magical' and obtained by operating only on the syntax of the relevant expressions (Steiner 1998, 163). However, such claims, both in general and in this particular case, are based on a representation of the relevant history that downplays or leaves out altogether the motivations in the relevant developments in physics together with the associated heuristic moves; once these are factored in, the discoveries start to look a lot less magical and mysterious (Kattau 2001).[15] In Dirac's case, a useful account of his essentially pragmatic attitude to the relevant mathematics can be found in Bueno (2005), in the context of a discussion of the infamous delta-function. Here Bueno is concerned with the dispensability of this function in the relevant physics but although his arguments do not, of course, apply to supposedly hybrid properties,[16] his analysis further supports the above point that paying attention to the relevant heuristic moves helps dispel any magical mystery mongering (see also Bueno and French forthcoming).

One could also dismiss the above as merely historical and of little relevance to how we should understand spin in today's context. Here it is the group-theoretic framework that underpins that understanding, but how one gets from the deployment of that framework to the conclusion that spin is a hybrid property remains unclear. Certainly there seems no obstacle to regarding that framework as representational or indexical in the manner suggested by Melia and Saatsi. Perhaps what we need is an appropriate understanding of properties in this context.

---

[14] A result that has been described as a 'turning point' in the field of microscopy (Mounce 2005).

[15] See also (French 2000) for a discussion of this point in the context of Wigner's famous concern about the applicability of mathematics.

[16] Not least because his argument depends in part on a distinction Azzouni draws between 'thick' and 'thin' epistemic access, where the criteria for the former all seem to be satisfied by spin (Azzouni 1997).

Morrison herself argues that the theoretical and experimental practices of physics have important implications for how we understand the notion of a physical property in general (op. cit., 548). This suggests that the argument from the role of group theory to hybridity is to be completed by a step involving a literal reading of these practices, to the effect that those aspects of the group theoretical description that — based on a broadly classical understanding of properties[17] — one might have been prepared to regard as 'merely' mathematical, or physically surplus, in some sense, should be regarded as describing a property of the world, but one that is both physical and mathematical. This also suggests that there is a particular view of physical properties acting as a foil, which must be rejected in the face of this literal reading. Although this view is not made explicit, statements such as the following suggest that it is one on which properties like spin, (rest) mass, charge etc. are regarded as monadic (and perhaps intrinsic):

> '... we typically think of the electron as being identified by properties like charge, mass and spin with each of these having a value for the single entity in question.' (548)[18]

The experimental problems with measuring spin together with its group theoretical description are then taken to imply that this view must be abandoned.

However, even if this is correct,[19] it does not immediately yield the conclusion that the property must be conceived of as hybrid. First of all, alternative views of properties are available that are more accommodating than the above. Thus, in the case of the kinds of properties that we are interested in when it comes to physics, one might take their group theoretic description as characterising their metaphysical nature.

For example, mass and spin are conceived of as eigenvalues of the operators of the Poincaré group (which is the group appropriate for the space-time of Special Relativity), where these operators label the irreducible representations of this group. These representations in turn yield differential equations, such as the Dirac equation for particles with these properties. Instead of focussing on these equations and how properties may then be conceived, we can shift our attention to the

---

[17] An understanding that might lead one to erroneously regard spin in terms of a particle actually spinning like a top in space-time.

[18] Earlier (ibid., 532) she refers to spin as an 'essential' property, although this is meant in an informal sense.

[19] We have already seen that the experimental issues are not decisive; although it could be argued that insofar as 'observations' of single spins involve a measurement context, they are not observations of spin qua quantum property.

group and the relevant representations. The result of combining the group transformations can then be presented in a 'multiplication' table, which allows various interesting features of the group to be displayed (such as whether it is abelian or not). Here, it is the inter-relationships between the group transformations that are fundamental and a property such as spin is effectively described in terms of what Eddington called a 'pattern of interrelatedness of relations'[20] (Eddington 1941, 278). This is not merely a colourful phrase but is indicative of the kind of metaphysics needed to accommodate this property ? certainly, one that moves beyond the typical focus on monadicity. Those metaphysical positions that accommodate such a view of fundamental physical properties as inter-related relations are likely to be more accommodating to the group-theoretic description of spin without having to ascribe any hybridity to it (see French forthcoming a).

Again, current trends in philosophy have not served us so well in this respect: as mentioned above, standard dispositionalist accounts of properties, in particular, appear to leave little room for these kinds of considerations. Nevertheless, alternatives are available, such as, again, Chakravartty's semi-realism. On this view, properties and relations are tied together in a holistic package that can capture the above 'inter-relatedness of relations'. Or, consider 'ontic' structural realism: according to this position,[21] objects and their intrinsic, monadic properties enjoy a secondary ontological status at best, with the underlying structure taken as ontologically primary.[22] This structure is then conceived of in terms of a web of inter-related relations that encompasses the relevant laws and symmetry principles represented group theoretically and again the above framework can be accommodated, with spin understood as one aspect of that structure.[23] Again, however, my point is not to advocate one such view over another here but simply to press the point that, first of all, some such view must be introduced and that, secondly, it must be appropriately metaphysically informed.

---

[20] Again, a structuralist metaphysics can help clarify what is meant here (cf. Ladyman and Ross 2007 on the metaphysics of 'patterns'). As I noted in fn. 11, the articulation of such a metaphysics is crucial if the structural realist is to avoid the accusation that she is restoring mathematics to a fundamental ontological role in this context.

[21] A form of which was developed by Eddington himself; see (French 2003).

[22] Objects are then either retained in a 'thin' form, where their identity is given in structural terms, or they are eliminated altogether depending on the variant of structural realism adopted (see French forthcoming a).

[23] Morrison herself considers and rejects the structural realist account; for a response see French forthcoming a).

12 Mixing in the Metaphysics

That an appropriate metaphysics of spin *qua* property is needed is further revealed by considering, in this context, the indexing/representational approach of Melia and Saatsi mentioned above in section 1. It might seem that, analogously to the use of physical rods of certain length to represent cicada life cycles, one could argue that we can express rotations, for example, via some kind of physical arrangement and in this case the relevant relation of combining elements might not even be that of actually performing successive rotations, obviating any need for a group-theoretic representation.[24] However there seems to be a straightforward response to this: when one considers the *representation* of the rotation group, the appropriate 'combining relation' is stated explicitly and if this relation were different, we would not be talking about the same group representation (see Eddington op. cit., 270). Of course, the defender of hybridity might take this as again indicative of the role of mathematics in understanding the relevant property. However, one can argue that it points to the significance of the relevant group transformations and their inter-relationships and the importance of constructing an appropriate metaphysical correlate.

Similarly, the 'weaseling' claim that the mathematics can be regarded as little more than heuristic scaffolding needs to be treated with some care. For sure, certain elements might appear to have only a heuristic role in this case: we can introduce the components of spin, for example, by specifying them in a set of mutually orthogonal planes (corresponding to spin in the x-, y- and z- directions) and after obtaining a group-multiplication table by taking the set of operations represented by rotations through 90° in each of the planes, we can effectively discard the planes and take this table (or the information it encodes) as representing the property (Eddington ibid., 279; see also French 2003). Here the planes perform only a heuristic role as scaffolding. However, in this case we are still left with the group table and so the defender of hybridity can argue that even with the scaffolding removed, the very fabric of the edifice so constructed is still interwoven with mathematics. Again, the response is to elaborate an appropriate metaphysics that will then allow us to 'detach' this mathematics as representational and eliminate any need to appeal to a hybrid nature.

One might object that there is no need for such a metaphysical account and that the physics itself gives us all the explication we need.[25]

---

[24]This is the suggestion made by Braithwaite in his critique of Eddington's philosophy of physics (French 2003).

[25]I am grateful to Margie Morrison for pressing me on this.

However, the articulation of the physics typically comes with an implicit metaphysics that, when probed, reveals an assumption to the effect that all fundamental properties are monadic. Consider charge for example: as characterised classically via Coulomb's Law, say, it is typically understood by philosophical commentators as a monadic property possessed by distinct entities, with the force law expressing the relationship between such properties. It is just such an understanding that encourages the intuition that we can straightforwardly conceive of a possible world in which there exists just one charged particle, for example, but this is an intuition that raises problems (see Haufe and Slater 2009; see also French and McKenzie forthcoming). Spin comes with no such understanding and it is, in part, my contention here that the failure to elaborate an appropriate metaphysical understanding opens the door to claims that it is some kind of hybrid or otherwise mysterious.

Furthermore, even the advocate of hybridity must appeal to some further metaphysical explication in order to resolve some of the problematic features of this notion. First of all, it is not clear how the mathematics can be constitutive of properties in this way, particularly if properties such as spin are taken to have causal impact and mathematics does not. Thus the idea would be that spin cannot be hybrid because it is causal and mathematical entities are not. Now introducing causality in this context is famously problematic and those who agree with Heitler and Morrison might well point out that it is precisely in this context – namely that of quantum physics – that their view gains some traction! Insisting that spin is 'physical' simply begs the question but noting that it can be related to observable phenomena via precisely the sorts of moves that Morrison herself so nicely sets out (cf. French and Ladyman 2003) may not be sufficient either. It could be argued, for example, that it is not necessary for a hybrid property to effectively manifest all aspects of its hybrid nature on every occasion.

Thus, one could insist that it is the 'physical' aspect of spin that features in such moves and is manifested at the observable level or that it is this aspect that participates in causal relations, such as those involved in the above measurements but that one still requires the mathematical aspect to make conceptual sense of the property, for example. At this point one might turn the argument around and suggest that the onus is now on the defender of the hybrid view to provide a fuller account of what it is to be a hybrid in this case, if the two aspects can be separated so cleanly. We might try to draw on an analogy with cases of hybridity elsewhere. Thus, various conceptions of 'hybrid objects' crop up across a range of academic disciplines, from computer science to history of science. They all seem to feature the conjunction of disparate features, from the use of both continuous and discrete variables in the former case, to interpretations that combine elements of the animal world and the mineral or the physical and the chemical in the latter. But in all cases, both sets of features are taken to do useful work and it is this that grounds the 'hybridity'. In the case of spin, with the physical aspect cleaved off, as suggested above, only the latter aspect would do any empirical work. Of course, as just indicated, the defender of hybridity could maintain that the mathematical aspect does another kind of work, either conceptual or explanatory. However, if the latter, then we bring in the points raised in the section 1 above, namely that an appropriate account of explanation then needs to be spelled out. If the former, then an opponent of this view can appeal to the kinds of metaphysical accounts that also and perhaps better do the same conceptual work and the mathematics would be reduced to having only an indexing or representational role again.

A more apposite example, perhaps, than those above is that of 'impure' sets, which can be taken to exist where their members do. So, for example, one might say that where we have two cicadas, we also have the number two. As is well known, this may allow one to maintain that numbers, sets, etc. should not be taken as abstract in the sense of not being spatio-temporally located but again, the point is that even if one were to grant this, it is the members of the set — the cicadas, for example — that wield any causal power and not the set — qua mathematical entity — itself (see Rosen 2001).

A further concern with ascribing causal power to spin, conceived as hybrid is that such ascriptions may be taken to underpin the identity of the properties concerned. But if spin is hybrid, its identity cannot be given by its causal powers alone, since its mathematical aspect cannot underpin or be associated with such powers. Again, of course, the issue of the role of causality in this context comes to the fore. A defender of the hybrid view might respond as follows:[26] insofar as the relevant causal powers are associated with or manifested via the kinds of moves Morrison spells out that relate the property to the empirical situation, they fall within the purview of and underpin the identity of, the physical aspect of the property. But insofar as they are associated with or articulated in terms of the association of instances of the property with other instances, or with different properties entirely, in a sense that is obviously vastly more attenuated than how we understand causality outside of this context, they underpin the identity of the

---

[26] Alternatively one could retreat to a quiddistic view of properties, although that is generally regarded as deeply problematic and would represent a high cost.

mathematical aspect. Thus we have a kind of 'double-identity' view that meshes with the hybrid nature of the property. Again, to respond, the opponent of the hybrid view needs to appeal to one of other of the available metaphysical packages in order to underpin the identity of spin. The structuralist package, for example, does so by conceiving of the inter-relationships both between instances of spin and between spin and other properties in terms of the relevant laws and symmetries, understood as features of the structure of the world (French forthcoming a). The point is, some such metaphysical account can be provided and that undercuts the motivation for ascribing hybridity in the first place.

A further option that might assuage this concern regarding identity conditions for properties would be to adapt Psillos's account of scientific models (forthcoming). Borrowing Dummett's phrase of 'physical abstract entities' to describe them, Psillos notes that models have physical properties ascribed to them; their identity (in part) is given by mathematical entities (such as phase spaces); and they are explanatorily relevant. The model of a simple harmonic oscillator, for example (ibid.), plays an important role both in explaining the behaviour of pendulums and in bringing under a single 'umbrella' a wide range of concrete entities; and it supports counterfactuals regarding, say, the results of changes to the length of a pendulum. Given these roles, and the commitment to such entities implied by reading theories literally, Psillos argues that causal inefficacy provides no grounds for denying reality to certain entities, and concludes: 'Causal inertia does not imply explanatory inertia.' (ibid.)[27]

Of course, one does not have to be a supporter of either the EIA or the hybridity claim to agree with that last point.[28] However, one can easily see how one might regard certain properties as 'physical abstract properties' in an analogous sense.[29] Thus, spin is obviously physical

---

[27]His aim in this work is not to defend the existence of mathematical objects but to argue that attempts to impose an 'austere' form of nominalist interpretation on scientific theories — in the sense that only concrete objects are admitted — would rule out many of those features of theories that give them their power.

[28]Psillos himself notes that if 'non-mathematical abstract objects' such as models are explanatory, so are the mathematical entities from which they are, in part, constituted. Of course, here the issue arises again of what counts as explanatorily relevant: granted that models feature in scientific explanations, one could argue that they still only play a representational role and cannot function as the actual explanandum. Thus, adopting Strevens' account, one might insist that the models, qua models, cannot be difference making.

[29]I would also reject the analogy, since I think it makes the mistake of reifying models. Just as in the case of theories, as indicated earlier, I adopt an eliminativist stance, arguing that although we can represent various aspects of scientific practice via (set-theoretic) models, for example, this does not licence us to claim that such models must

in at least certain respects, plays crucial explanatory and unificatory roles and, significantly, one can argue that its identity as a property is given in group theoretical terms. However, it is then a further step to claim that this amounts to the identity conditions for the property being mathematically grounded.[30] One could, for example, insist again that group theory is acting in a representational capacity here and that it is the underlying structure — conceived of from the perspective of either structural realism or semi-realism, or indeed, some other ontological stance — that ultimately grounds the identity of this property. Of course, the availability of such alternatives does not provide a knock-down argument against mathematics performing such an identity grounding role in this case and the supporter of the hybridity claim might wonder where the argumentative onus actually lies here. Nevertheless, one can see how the costs of the hybrid view begin to mount up.

Thus consider, as a final comment, the point that, on such an account, properties could not be taken to be 'firmly rooted' in the spatio-temporal world as some views have it (Swoyer 2000) but, depending on one's philosophy of mathematics perhaps, would have to be regarded as 'transcendental' and acausal or 'other-worldly' themselves. In this case, standard philosophical accounts of property instantiation will have to be revised to allow for the role of the mathematics. Of course, such requirements are not impossible to satisfy;[31] my point here is just to indicate some of the costs associated with the hybridity claim. And given that there are alternative views of properties that may be more accommodating to the features associated with the group-theoretic description of spin, I suggest that as in the case of the Exclusion Principle, this move to an ontological construal of the role of mathematics can

---

exist, whether as 'physical abstract entities' or any other kind of entity. And in case anyone wheels on stage physical models such as Crick and Watson's model of DNA structure, let me say that I am an eliminativist about macroscopic objects as well! But that's a discussion for another time (see French forthcoming a).

[30] Note how this is different from the case of models, under Psillos' conception: as he notes, if these are considered to be abstract then there is little obstacle to their identity being partly constituted by other abstract objects. In the case of a (at least partly) physical property such as spin, one again bumps up against the issue of hybridity: how do the mathematically grounded identity conditions relate to the physically grounded ones in such a way as to fix the identity of spin as a property?

[31] One might adopt a partial instantiation account for example: when a property is instantiated only its physical aspect is instantiated such that it can be said to be 'rooted' in space-time. Since instantiation appears mysterious anyway — or, more politely perhaps, is claimed to be *sui generis* — I have no doubt that it is flexible enough to accommodate all kinds of modifications (for critical discussion of standard accounts of instantiation, see Vallicella 2002).

also be blocked.

13 Conclusion

I hope to have shown two things: first, that when it comes to the EIA the possibility of mathematical entities acquiring a non-causal but explanatory role is not well motivated, even within the framework of an account of explanation that might be sympathetic to such a role. Secondly, that in the case of spin, the assertion of hybridity also lacks strong motivation and comes with associated metaphysical costs.

The latter case can be taken to exemplify such assertions with regard to the kinds of properties one encounters in modern physics, given the role of group theory in affording us an appropriate conceptual grasp on the property and its widespread, if not ubiquitous, presence in the field. Spin offers perhaps the most striking case given, as already mentioned, the well-known problems in conceiving of it in classical terms, but given how charge is also presented in the quantum/relativistic context, one could also mount a hybridity claim in this case as well, which will be similarly undermined by considerations such as those presented above.[32]

In both the case of the EIA and spin we can disentangle the descriptive role of mathematics from the explanatory role of the physical. However, insofar as a non-question begging account of explanation can be given it may be difficult to insist on the mathematics having only an indexing or representational role. What is needed to achieve the disentangling is to provide an appropriate metaphysics of symmetry and 'hybridity' respectively. I suggest, then, that the debate needs to move on from discussions of explanation and representation to the consideration of such a metaphysics.

Acknowledgements

Versions of this paper was given at the 'Mathematical and Geometrical Explanation in Physics' conference, Bristol University, December 2009, the Annual Conference of the British Society for the Philosophy of Science, University College Dublin, July 2010 and 'The Role of Mathematics in Science' workshop at the University of Toronto, October 2011. I'd like to thank Alan Baker, Anjan Chakravartty, James Ladyman, Kerry McKenzie, Margie Morrison, Ioan Muntean, Christopher Pincock and

---

[32]It may be that the claims of Franklin (1989) and Parsons (2008) with regard to 'quasi-concrete' objects might be similarly undermined but I shall leave that to a future work. I am grateful to a reader of a previous version for reminding me of this work.

Juha Saatsi for helpful comments and encouragement. The responsibility for its content is entirely mine. This work was initially supported by a Major Research Scholarship from the Leverhulme Trust.

BIBLIOGRAPHY

[1] J. Azzouni, Thick epistemic access: Distinguishing the mathematical from the empirical, *Journal of Philosophy* 94: 472-484, 1997.
[2] A. Baker, Does the Existence of Mathematical Objects Make a Difference, *Australasian Journal of Philosophy* 81: 246-264, 2003.
[3] A. Baker, Are There Genuine Mathematical Explanations of Physical Facts?, *Mind* 114: 223-238, 2005.
[4] A. Baker, Mathematical Explanation in Science, *British Journal for the Philosophy of Science* 60: 611-633, 2009.
[5] R. W. Batterman, (2010), On the Explanatory Role of Mathematics in Empirical Science, *British Journal for the Philosophy of Science* 61: 1-25, 2010.
[6] M. Berry and J. M., Indistinguishability for quantum particles: spin, statistics and the geometric phase, *Proc. R. Soc. London*, Ser. A 453: 1771-1790, 1997.
[7] A. Bird, *Nature's Metaphysics: Laws and Properties*. Oxford: Oxford University Press, 2007.
[8] K. Brading and H. Brown, Symmetries and Noether's Theorems. In K. Brading and E. Castellani (eds.), *Symmetries in Physics: Philosophical Reflections*. Cambridge: Cambridge University Press, 89-109, 2003.
[9] K. Brading and E. Castellani (eds.), *Symmetries in Physics: Philosophical Reflections*. Cambridge: Cambridge University Press, 2003.
[10] O. Bueno, Dirac and the dispensability of mathematics, *Studies in History and Philosophy of Modern Physics* 36: 465-490, 2005.
[11] O. Bueno and S. French, Can Mathematics Explain Physical Phenomena?, *British Journal for the Philosophy of Science* 63 (1): 85-113, 2012.
[12] O. Bueno, S. French and J. Ladyman, On Representing the Relationship between the Mathematical and the Empirical, *Philosophy of Science* 69: 452-473, 2002.
[13] O. Bueno and S. French (forthcoming), From Weyl to von Neumann: An Analysis of the Application of Mathematics to Quantum Mechanics.
[14] A. Chakravartty, *A Metaphysics for Scientific Realism*. Cambridge: Cambridge University Press.
[15] M. Colyvan (forthcoming), Causal Explanation and Ontological Commitment. (available at: homepage.mac.com/mcolyvan/papers/ceaoc.pdf)
[16] F. Correia, Ontological Dependence, *Philosophy Compass* 3/5: 1013-1032, 2008.
[17] N. C. da Costa and S. French, *Science and Partial Truth*. Oxford: Oxford University Press, 2003.
[18] M. Dorato and L. Felline (forthcoming), Scientific Explanation and Scientific Structuralism.
[19] A. Eddington, Discussion: Group Structure in Physical Science. *Mind* 50: 268-279,1941.
[20] J. Franklin, Mathematical Necessity and Reality, *Australasian Journal of Philosophy* 67: 286-294, 1989.
[21] S. French, S., The Reasonable Effectiveness of Mathematics: Partial Structures and the Application of Group Theory to Physics, *Synthese* 125: 103-120, 2000.
[22] S. French, S., Scribbling on the Blank Sheet: Eddington's Structuralist Conception of Objects, *Studies in History and Philosophy of Modern Physics* 34: 227-259, 2003.
[23] S. French, S., Symmetry, Invariance and Reference. In M. Frauchiger and W. K. Essler (eds.), *Representation, Evidence, and Justification: Themes from Suppes*. (Lauener Library of Analytical Philosophy; vol. 1) Frankfurt: Ontos Verlag, 127-156, 2008.
[24] S. French, S.,Keeping Quiet On the Ontology of Models, *Synthese* 172: 231-249, 2010.
[25] S. French (forthcoming a), The Structure of the World: From Representation to Reality
[26] French, S. (forthcoming b), Semi-realism, Sociability and Structure, forthcoming in *Erkenntnis*.
[27] S. French and D. Krause, *Identity in Physics*. Oxford: Oxford University Press, 2006.
[28] S. French and J. Ladyman, J., 'Remodelling Structural Realism: Quantum Physics and the Metaphysics of Structure: A Reply to Cao', *Synthese* 136: 31-56, 2003.
[29] S. French, S. and K. McKenzie, Thinking Outside the (Tool)Box: Towards a More Productive Engagement Between Metaphysics and Philosophy of Physics, *European Journal of Analytic Philosophy* 8: 42-59, 2012.
[30] S. French, Steven and D.Rickles, Understanding Permutation Symmetry. In K. Brading and E. Castellani (eds), *Symmetries in Physics: New Reflections*. Cambridge: Cambridge University Press, 212-238, 2003.

[31] M. Friedman, M., Explanation and Scientific Understanding, *Journal of Philosophy* 71: 5-19, 1974.
[32] A. Guay and B. Hepburn, Symmetry and Its Formalisms: Mathematical Aspects, *Philosophy of Science* 76: 160-178, 2009.
[33] H. Halvorson M. Müger, M., Algebraic quantum field theory. In J. Butterfield and J. Earman (eds.), *Handbook of the philosophy of physics*. Kluwer, 731-922, 2006.
[34] C. Haufe and M. H. Slater, Where No Mind Has Gone Before: Exploring Laws in Distant and Lonely Worlds, *International Studies in the Philosophy of Science* 23: 26-276, 2009.
[35] W. Heiter, *Man and Science.*, London: Oliver and Boyd, 1963.
[36] S. Kattau, Kabbalistic Philosophy of Science: Review of The Applicability of Mathematics as a Philosophical Problem, by Mark Steiner, *Metascience* 10: 22-31, 2001.
[37] P. Kitcher, P., Explanatory Unification and the Causal Structure of the World. In P. Kitcher and W. Salmon (eds.), *Scientific Explanation*. Minneapolis: University of Minnesota Press: 410-505, 1989.
[38] J. Ladyman and D. Ross, *Everything Must Go*. Oxford: Oxford University Press, 2007.
[39] D. Lewis, Causal Explanation. *Philosophical Papers*, Volume II. 214-40. Oxford: Oxford University Press, 1986.
[40] M. Massimi, *Pauli's Exclusion Principle: The Origin and Validation of a Scientific Principle*. Cambridge: Cambridge University Press, 2005.
[41] J. Melia, Weaseling Away the Indispensability Argument. *Mind* 109: 458-79, 2000.
[42] R. Mirman, R., *Group Theory: An Intuitive Approach*. Singapore: World Scientific, 1995.
[43] M. Morrison, Spin: All is Not What it Seems, *Studies in History and Philosophy of Modern Physics* 38: 529-557, 2007.
[44] D. Mounce, D., Magnetic resonance force microscopy, *Instrumentation & Measurement Magazine*. IEEE, 8: 20-26, 2005
[45] W. J. Mullin and G. Blaylock, G., Quantum statistics: Is there an effective fermion repulsion or boson attraction?, *American Journal of Physics* 71: 122-1231, 2003.
[46] C. Parsons, *Mathematical Thought and Its Objects*. Cambridge: Cambridge University Press, 2008.
[47] C. Pincock (forthcoming), Mathematical Contributions to Scientific Explanation.
[48] S. Psillos (forthcoming), Scientific Realism: Between Platonism and Nominalism.
[49] G. Rosen, G., Abstract Objects, *Stanford Encyclopaedia of Philosophy*, http://plato.stanford.edu/entries/abstract-objects/, 2001
[50] D. Rugar, R. Budakian, H. J. Mamin & B. W. Chui, Single spin detection by magnetic resonance force microscopy, *Nature* 430: 329-332, 2004.
[51] J. Saatsi, Discussion: The Enhanced Indispensability Argument: Representational vs. Explanatory Role of Mathematics in Science, *British Journal for the Philosophy of Science* 62: 143-154, 2011.
[52] D. Shapere, Notes Towards a Post-Positivistic Interpretation of Science. In P. Achinstein and S.F. Barker (eds.), *The Legacy of Logical Positivism*. The Johns Hopkins Press, 115-160, 1969.
[53] M. Steiner, *The Applicability of Mathematics as a Philosophical Problem*. Cambridge, Massachusetts: Harvard University Press.
[54] M. Strevens, *Depth: An Account of Scientific Explanation*. Harvard University Press, 2008.
[55] E. C. G. Sudarshan and I. M. Duck, I.M., What price the spin-statistics theorem?, *Pramana*. Indian Academy of Sciences, 61: 645-653, 2003.
[56] P. Suppes P., What is a Scientific Theory?. In S. Morgenbesser (ed.) *Science Today*. New York: Basic Book; Inc.: 55-67, 1967.
[57] C. Swoyer, Properties, *Stanford Encyclopaedia of Philosophy*, http://plato.stanford.edu/entries/properties/, 2000.
[58] W. F. Vallicella, Relations, Monism, and the Vindication of Bradley's Regress, *Dialectica*, 56: 3-35, 2002.
[59] B. van Fraassen, *Scientific Representation: Paradoxes of Perspective*. Oxford University Press.

Environment, Action Space and Quality of Life:
An Attempt for Conceptual Clarification

SILVIA HARING AND PAUL WEINGARTNER

ABSTRACT. The paper offers a conceptual clarification of the terms environment, action space and quality of life. Although these terms are used quite frequently in different scientific domains, they are not very well clarified and therefore lead to several confusions in discussions about quality of life. The paper begins with a definition of environment (first general, then as physical, biological and mental environment) which is a basic concept presupposed by several other important concepts used in human sciences; it continues with definitions of active and passive action space to be followed by definitions for quality of life on different levels and its presupposed values. Finally a clarification of lack and loss of quality of life is given.

1 Introduction

The terms "environment", "living space", "action space",[1] "quality of life" ("well-being") occur in many scientific treatments of different domains like sociology, psychology, medicine, geography, environmental research and philosophy. Despite some exceptions[2] these terms are not sufficiently clarified and are used only on the basis of their meaning in everyday language. Insufficient conceptual clarification leads to ambiguities and consequently amounts to a barrier for comparing the results in different domains or disciplines.

It is a great pleasure and honour for us to be able to contribute this essay to the volume to honour Pat's 90th birthday. Pat Suppes has always been a master of conceptual clarification. His two books *Introduction to Logic* and *Axiomatic Set Theory* are masterpieces in this respect. One of us remembers his first lectures in Salzburg in 1974 at the Summer

---

[1]The respective German terms "Lebensraum" and "Lebenswelt" had a special political meaning in National Socialism (between 1933 and 1945). This special meaning does not concern us here. The definitions given in this essay are independent of it, like those of well-known sociologists and psychologists. Cf. Lewin (1951) p.99.

[2]Cf. Weichhart (1979), Bunge (1979) and Mahner–Bunge (1997).

School of the University of Salzburg and the Institut fuer Wissenschaftstheorie which showed his great didactic skills. Further examples are Pat's clarifying contributions in the domain of linguistics. Among them there is the essay on *Congruence of Meaning* which didn't get the attention it would have deserved and the new ideas of which should be further developed. And even in a field in which we find lots of confusion — metaphysics — he successfully applied his conceptual clarification by introducing probability. Taking Pat as model we hope to achieve a respective valuable clarification of "environment", "action space" and "quality of life" in this essay.

1.1 A first task of conceptual clarification is to find out which concepts are presupposed by which others. It will turn out that a conceptual clarification of *living space* (and *action space*) and of *quality of life* presupposes a clarification of environment. In a similar way *well-being* presupposes a determination of *quality of life*. Concerning *living space* there are two possibilities: we can characterise it as a part of the environment and independently of the *quality of life* or as that part of the environment which promotes or which hinders the *quality of life*. Concerning the second possibility a clarification of *living space* presupposes that of *quality of life*. We shall discuss therefore our conceptual explications in the following order: Environment, living space, quality of life, well-being and lack of quality of life.

1.2 A further task of this essay is to make some suitable restrictions. A first restriction concerns the individuals of whom the environment and their quality of life is investigated. These are first of all human persons and secondly communities (like families) and societies. We shall not investigate environment, living space and quality of life of animals or plants. A second restriction is concerned with the terms environment and living space. For example "geographical environment" would be too wide, since it contains parts which have no relevant relation to the person whose environment it is.

1.3 In contradistinction to a clarification of environment for that of quality of life it will be necessary to introduce values and goals. One cannot speak of quality of life if basic values like survival are not realized.

In this connection it is also important to deal with losses of quality of life. They may be of different kind, for example those which are imposed to humans by their environment or those that are accepted in order to reach a higher quality of life (for example if an operation is accepted in order to regain health).

2 Environment

As mentioned already in 1.2 it does not seem reasonable to determine the term "environment of person x" by either the physical or chemical or biological environment of x. Since such a demarcation is too wide because each domain contains things which stand in no relevant relation to x. Moreover it would be too narrow if one of the above domains is taken. Also an explication with the help of geographical environment or of topological environment suffers from the same disadvantage to be too wide.[3] On the other hand these too wide demarcations seem to be too narrow in another sense: if we want to say that also concepts, hypothesis, theorems or the "content" of a book[4] belong to our environment then these things are not included in any of the above mentioned domains. And further if we want to say that also objects like perceptions, imaginations, thoughts, desires, feelings, free will decisions belong to our environment and understand them as *mental* objects then they are also not included in the above domains. Other classifications consider the functions of the environment. Thus Dunlap–Catton[5] distinguish the following three functions of the environment: supply depot (resources necessary for life like air, water, shelter ...); waste repository (man produces throw away products and puts them into the environment), living space (where humans live, work, play, travel ... in general the earth). We do not use this classification since it is neither disjoint nor complete: It is not disjoint because there are many things which belong to both supply depot and living space. It is not complete because conceptual objects like concepts, hypotheses, theories ... (see below) do not belong to either class.

In order to meet these restrictions we understand by "environment" of a person x (community, society) only those things which stand in the relation of mutual interaction (understood in a wide sense) to x.[6]

Finally, we want to mention that we do not define "person" ("human person"), "community" or "society" in this essay, although these expressions are used in the subsequent definitions. To define these terms

---

[3]Cf. Weichhart (1979).

[4]In Popper's terminology these are objects of his "third world". These objects are conceptual and also non-material in the sense of "unembodied" in Popper's terminology (Cf. Popper–Eccles 1977. ch. P2).

[5]Dunlap–Catton (2005), p. 242 f.

[6]The following definition is analogous to that of Bunge (1979) p. 7. However there might be some difference. Whereas Bunge accepts concrete material objects and conceptual objects we allow in addition non-material mental objects as members of our environment. D1 is also similar to the definition of environment by Friedrichs (1950). Cf. Weichhart (1979) p. 525.

is not the task of this article; there are extensive studies of these concepts in papers and books. We are therefore presupposing here the usual understanding of these terms in recent works of social psychology.

DEFINITION 1 (D1). Let x be a human person (a community of persons, or society).
Then the *environment* of x consists of those things y $(x \neq y)$ such that

(a) y acts on x (or on a part of x)

(b) x (or a part of x) acts on y

Examples: Things on our desk which we use all the time, the members of our families, parts of our society, our flats or houses, our climate, last but not least also the thoughts, ideas, desires of others and conceptual entities like hypotheses, theories, world views, etc.

2.1 Classification of things of the human environment In order to be more accurate concerning the objects belonging to the environment of a human person or community or society we propose the following classification.

NO Natural objects without consciousness: stones, mountains, lakes, plants,[7] lower-level animals,[8] etc. Concrete properties or states of concrete NOs will be classified as NOs. For example a concrete heat radiation of the sun on an object will be counted as a NO

NB Natural objects (animals) with consciousness[9]

NB- NBs without rationality: higher-level animals

NB+ NBs with rationality: humans

---

[7] According to Aristotle and to many thinkers in the Middle Ages all living beings have a "soul". The "soul" is the principle of life which realises the three essential properties of a specific living being: nutrition, growth and propagation. These three properties are still involved in the biological concepts presently used: metabolism includes the first two and reproduction the last. The gene-identity was more roughly expressed by the uniqueness of the individual substance.

[8] Lower-level animals have stimulus-response reaction like plants (for example phototropism) and in addition some grasping of objects but not real perception. Cf. Mahner–Bunge (1997) for this demarcation.

[9] For consciousness neither stimulus-response reactions nor some grasping of objects are sufficient. Sensation and some kind of inner experience and subjective experience of one's own inner states would be sufficient to speak of consciousness. One need not require having self-consciousness for higher animals who possess consciousness.

KA Concrete artefacts. Concrete artefacts are produced by humans out of natural objects (or parts of it)
Examples: individual concrete computers, microscopes, instruments, houses, roads, highways, railways, cars, etc. Observe that "computer", "electron-microscope", "house" are generally not used as expressions for concrete individual objects; their usual meaning is a conceptual object containing the respective peculiar characteristics of the referred *class* of objects.
Since we have required that KAs are produced by humans, we shall classify the nest of a bird or the cave of a fox under natural objects.

CO Conceptual objects
Examples: concepts, hypotheses, theories, meanings, extensions, intensions, numbers, inferences, arguments, probabilities, propositions, etc.[10] Observe that individual concrete sentences (tokens) are KAs, but sentences as types (i.e. classes of sequences of signs of the same form) are COs.

SO Supernatural objects
Examples: angels, God. Supernatural objects belong to the environment of religious people

2.2 Physical/biological and mental environment

DEFINITION 2 (D2). Let x be a human person, a community or a society. Then the *physical/biological environment* of x are those things y (where $x \neq y$) such that:

1. D1 is satisfied

2. y belongs to NO or y belongs to KA

We agree with Karl Popper and John Eccles that consciousness and rationality are not reducible to brain processes.[11] For short: Mind is not reducible to body. On the other hand NBs, NB-s and NB+s are matter dependent living things (animals and humans). Therefore we also accept an interactionist theory of the relation between mind and body. That means also that we accept an interaction between mental processes like perceiving, thinking, desiring, intending, feeling, willing and bodily movements mediated by brain processes.

---

[10]Cf. Popper–Eccles (1977) P2.

[11]Cf. Popper–Eccles (1977) chs. P3 and E7. This however is not the place to discuss the ramifications of the perennial mind-body problem.

However we cannot go into this topic more deeply because this is not an article on the mind-body problem. It is just that we need some rough demarcation in order to define some further concepts (action space, quality of life). We do not count mental processes as things of the environment, since they do not act themselves but belong to human persons or higher animals; only the latter act on things of their environment even if mental processes are causal factors for such actions. But since human persons possess such mental actions we say that they are both physical/biological *and* mental individuals. Therefore they can also have a respective environment. And this holds also in a more restricted sense of animals with consciousness but without rationality.

DEFINITION 3 (D3). Let x be a human person. Then the *physical/biological-mental environment* of x are those things y ($x \neq y$) such that:

1. D1 is satisfied

2. y belongs to NB, to NB- or NB+

In an analogous way we might also define *conceptual environment* and *supernatural environment*.

2.3 Possible interactions among the things of the environment

It is not the topic of this paper to give a more accurate analysis of the many possible interactions among the things of the environment. Therefore we restrict ourselves to a rough classification of interactions which is needed for the subsequent chapters of the essay. In order to avoid confusion we first point out that sets (classes) of objects cannot act on other sets (classes) of objects, because sets (classes) are conceptual objects. In order not to say that NO acts on NO (where NO is understood as a class of objects) we take the terminology to speak of NOs in the plural, i.e. meaning several singular natural objects (similar for other objects of the environment).

1. NOs act on NOs in accordance with physical, chemical or biological laws.

2. NOs act on NBs in accordance with physical, chemical or biological laws and this effect can be experienced consciously in NBs and can be treated rationally in NB+s.

3. NOs act on KAs in accordance with physical, chemical and biological laws.

4. NOs cannot directly act on COs and in no way on SOs. However since NOs can act on NB+s who can produce COs with the help of their mental processes we may say that NOs can act indirectly on COs. For example the observation of sun-spots (by scientists, NB+s) can be a motive (for scientists) for proposing hypotheses (COs) about changes of climate. Similarly on a lower level NOs can act on NB-s who may produce some sort of hypothesis (CO) with the help of their processes of consciousness.[12]

5. KAs act on NOs, NBs and KAs in accordance with physical, chemical and biological laws.

6. KAs act via processes of consciousness and via mental processes of NBs and NB+s on NOs, KAs, NBs and NB+s. For example defects on computers motivate scientists to improve them.

7. NB-s act on NOs. For example higher animals (lions, zebras) can act on waters, plants and animals without consciousness.

8. NB-s act on NB-s and NB+s. For example higher animals can act on higher animals and on humans.

9. NB-s act on KAs. For example higher animals might destroy fences and shelters.

10. Can NB-s act on COs or produce COs. This is a difficult question. According to Popper and Eccles all animals with consciousness may have hypotheses in a primitive sense. At least in the sense that learning via trial and error uses a hypothesis (trial) which leads either to success or if not will be rejected and replaced by a new one.[13]

11. Humans (NB+s) act on NOs, NB-s, KAs, NB+s (i.e. on other humans) first of all according to physical, chemical and biological laws. Secondly humans act on NOs, NB-s, KAs, NB+s and COs with the help of their consciousness and rationality via their mental processes.

12. COs cannot act directly on NOs or KAs. This is only possible in an indirect way via NBs and NB+s and their conscious and mental processes. For example a theory for building railway bridges

---

[12]Popper and Eccles (1977) p. 128 defend that animals with consciousness are "active problem solving agents ... attempting to control its environment". In this sense can NOs indirectly act on COs (Cf. ch. P2).

[13]Popper and Eccles (1977) p. 134ff.

(COs) may act via engineers and their mental processes (NB+s) on the constructed bridges (KAs). Scientific results (COs) guide technicians to improve landscapes, medical instruments, etc.

13. According to religious belief (faith) supernatural objects can act on any objects either directly or indirectly via humans and their mental processes.

According to this list of mutual interactions we may say now more accurately which things belong to the environment of a human person. A more refined definition of environment is therefore as follows:

DEFINITION 4 (D5). Let x be a human person.[14] Then the *environment* of x are those things y ($x \neq y$), such that the conditions (a) - (d) are satisfied:

(a) y is a natural object without consciousness (a member of NO) and acts on x (cf. 2.) or x acts on y according to laws of nature and with the help of their consciousness and rationality via their mental processes (cf. 11).

(b) y is a natural object with consciousness (a member of NB with or without rationality) and acts on x (cf. 8., 11.) or x acts with the help of consciousness and rationality via its mental processes on y (cf. 11)

(c) y is a concrete artefact (a member of KA) and acts on x (cf. 5.) and this effect can be experienced consciously and treated rationally by x (cf. 6.) or x acts with the help of consciousness and rationality via its mental processes on y (cf. 11.).

(d) y is a conceptual object (a member of CO) and acts via consciousness and rationality and their mental processes on x (cf. 12.) or x acts with the help of consciousness and rationality and its mental processes on y (cf. 11.).

3 Living Space

By "living space" or "action space" of a person or a community or society we understand a part of the environment. Such a part can be selected in different ways. We propose two selections, the first of which is a spatial and causal demarcation the second is obtained by the question

---

[14]We didn't add here community or society since we cannot speak of the consciousness, rationality or of the mental processes of a community or society (but only in reference to individual persons).

whether this part contributes positively or negatively to the quality of life.

Concerning the first selection we distinguish between active and passive action space and then will determine living space as both active and passive action space:

DEFINITION 5 (D6). Active Action Space

Let x be a human person (or community or society). Then the *active action space* of x are those things y ($x \neq y$) such that: x or a part of x acts on y.

According to this definition the active action space embraces all those things of the environment which are acted upon or changed by human persons, communities or societies. For example houses, streets, hospitals, technical instruments, agricultural products, but also pollution, CO_2 surplus ... etc.[15]

DEFINITION 6 (D7). Passive Action Space

Let x be a human person (or ... etc.). Then the *passive action space* of x are those things y ($x \neq y$) such that: y acts on x or on a part of x.

We shall define living space as active and passive action space:

DEFINITION 7 (D8). Living Space 1

Let x be a human person (or ... etc.). Then the *living space* of x are those things y ($x \neq y$) such that both:

1. y acts on x (or on a part of x)

2. x (or a part of x) acts on y

Observe that living space $\neq$ environment since living space is the intersection of active and passive action space whereas environment (D1) is the union of both. Thus living space 1 $\subset$ environment.

Living space could be restricted further to those things of the environment which can be reached by our senses (sense organs) + their technical extensions (for example microscopes and telescopes) and which are effective on our senses (+ extensions).

There is another plausible demarcation of living space which relates to quality of life and which we call living space 2. The criterion for this demarcation is its contribution to the quality of life. This contribution may be positive or negative. Thus clean air and polluted air, sufficient healthy food, or sufficient medical care belong to living space 2. On the other hand those kinds of radiation which act causally on us but do not

---

[15]Cf. Weichhart (1979) p. 525. This demarcation and the one given by D8 goes back to Uexkuell (1909).

contribute positively or negatively to our quality of life do not belong to our living space 2 although they belong to our environment according to D1. This is the case for example of the cosmic background radiation or of those parts of the electromagnetic spectrum which is outside our visual domain or those items of media information which do not contribute to our quality of life. Living space 2 we might therefore define as follows:

DEFINITION 8 (D9). Let x be a human person (or ... etc.). Then the *living space 2* of such x are those things y ($x \neq y$) of the environment of x (D1) such that:

1. y contributes positively to the quality of life of x or

2. y contributes negatively to the quality of life of x

4 Quality of Life (QL)

The expression "quality" need not be interpreted as involving values. For example if "quality" is the opposite of "quantity" then both are value-neutral. Also, if quality is one of Aristotle's 10 categories, then it is not value-laden. However if "quality" occurs in contexts like "quality of life" (German: "Lebensqualität"; French: "qualitee de la vie") then "quality" is certainly not value-free, but value-laden. But which kinds of values are involved here? This question shows that QL seems to be relatively dependent on a certain type of values or on a certain level of values. Thus a basic level of QL will correspond to basic values and a higher level of QL will correspond to higher values. This leads first of all to a very general definition of QL:

DEFINITION 9 (D10). Let x be a human person (or ... etc.). Then the quality of life (QL) of x is a certain level of realisation of values or goals of x within the life-time (or part of it) of x.

According to D10 QL of x can be understood in a twofold

(i) as a state of x at time t

(ii) as a property of a period of life of x

4.1 Basic level of quality of life

In connection with definition D10 there is the question whether there is a kind of basic level of realisation of values or goals which has to be satisfied for any type of quality of life. We think that there is. And moreover we think that this basic level of quality of life depends on basic values, especially on the value of survival and on the value of health. Therefore we begin with the definition of basic value.[16]

DEFINITION 10 (D11). Let x be a human person (or ... etc.). Then v is a basic value for x iff

1. meeting v is necessary for x to stay alive in the environment of x
or
2. meeting v is necessary to keep or regain health for x in the environment of x

Examples: clean air, clean water, adequate food, shelter, clothing ... etc.

Conditions 1. and 2. seem to be universally valid for living organisms where all its main parts contribute to satisfy this goal:

> All the various biological processes that occur along these five levels of organisation (cells, tissues, organs, organ systems, organisms) can be seen to serve the life of the organism as a whole.[17]

With the help of definition D11 we may now define *basic level (bl) of quality of life (QL)*.

DEFINITION 11 (D12). Let x be a human person (or ... etc.). Then bl is a *basic level* of QL for x iff bl is the realisation of basic values for x

As has been mentioned (D10) basic level of QL can be understood either as a state of x at time t or as a property of a period of life of x.

Besides the basic values there are a number of valuable states of affairs which support the basic values. For example growing up and safety in a family, belonging to a community, possibility for labour, recreation, sport ... etc. These we may call *supporting values*.

---

[16]Definition 11 is originally due to Bunge (1989) p. 35. Concerning the concepts of *value* and *basic value* we are presupposing a rough understanding. An analysis of values is not the task of this article. If we say that the survival (or staying alive) of x or keeping or regaining (the preservation of) health of x are (basic) values for x then the value (of x) is a property of x. If we say that clean air, clean water, unadulterated food ... are necessary for the preservation of health then these things — or better the respective properties cleanness of water ... — are understood as means or mean-values for satisfying the basic values. Sometimes also states or states of affairs are interpreted as values: for example if we say that it is an important value for x to freely choose his (her) partner or to be able to preserve his health for a longer period of time. For the following considerations we leave it open whether a thing, a property or a state of affairs is said to be a value, since it will be clear from the context. For an accurate analysis of values see Bunge (1989) and Weingartner (1996).

[17]Lofti (2010) p. 129.

A further point worth mentioning is that there are pairs of mutually corresponding values where one of them belongs to the human person, the other belongs to the environment. For example: clear air and healthy lungs, adequate food and healthy digestive system, rough climate and inurement of the body ... etc.

With respect to the basic values defined in D11 we want to stress that they are not understood as "absolute". We think that it holds as a statistical law that humans have a natural inclination towards the realisation of basic values. Statistical laws allow exceptions. And there are some exceptions in extreme situations: hunger strikes, martyrdom, acts of terrorism.

In this case basic values are sacrificed for a kind of ultimate or final value or goal. This is a type of value which is invariant w.r.t. other human desires and goals. According to Aristotle this ultimate goal is happiness. With respect to happiness all other goals are subordinated. Nevertheless it holds first that the basic values are those that are *naturally desired* by all humans, which means that they are not subject to free will decisions. Secondly, in most cases they are necessary for man to be happy; i.e. they are usually also necessary means for happiness or for the ultimate goal.

4.2. Mental Quality of Life

There is the question whether there exist basic values in a spiritual or mental sense analogous to the basic values defined by D11, i.e. survival and health. Is there a kind of mental survival or mental health? We may approach that question by asking if there are cases where mental survival or mental health is in danger to break down. For example in case of a permanent deep depression or when a person possessing physical (bodily) health intends to commit suicide. These examples seem to show that in certain cases there is danger to lose mental survival and mental health. Therefore it seems justified to consider conditions and if possible necessary conditions which could hinder such a breakdown.

DEFINITION 12 (D13). Let x be a human person. Then v is a *mental basic value* for x iff

1. meeting v is necessary for x to mentally survive or

2. meeting v is necessary for the mental health of x

There are different theories which defend specific mental basic values. One is the logotherapy of Victor Frankl. According to this theory at least one of the following conditions are necessary for mental survival and mental health: a task or a personal you.[18] A task which holds a person's mind enthralled (be it a profession or a task beside the profession) keeps him (her) in mental health and mental survival. Similarly if there is a beloved person who loves you. He (she) who has both is happy according to Frankl. An extreme situation of bodily and mental survival is described by Frankl (2005) when he was hanging on for dear life in the concentration camp.

According to D13 we may define a basic level of mental quality of life as follows:

DEFINITION 13 (D14). Let x be a human person. Then ml is a *basic level of mental quality of life* for person x iff ml is the realisation of mental basic values for x

Relying on this basic level of mental quality of life higher levels of quality of life will be described subsequently.

4.3 Higher Level of Quality of Life

One type of values which belong to a higher level of QL are those which are concerned with the society to which the person or community belongs. Such values are dependent on the rights and duties of the respective social environment. But there are some which seem to be invariant w.r.t. many societies. One important type concerns values which are legitimate and accessible for the person belonging to a society.

DEFINITION 14 (D15). v is a *higher value* for the human person x who belongs to society g iff

1. v is desired (or viewed as desirable) by all or statistically most members of g

2. v is legitimate in g

3. v can be realised by all or statistically most members of g

4. for some v, the realisation of v by members of g may be obligatory

Examples: Increasing knowledge according to ability and interest, improving control about natural forces, human life in peace, average welfare, sufficient medical treatment, education, school attendance.

---

[18]Frankl (2004) p. 101 ff., p 166 ff., p. 178 ff.

DEFINITION 15 (D16). v is legitimate for person x living in society g iff v can be met by x in g

1. without hindering the satisfaction of any basic value or need of any member of g

2. without endangering the integrity of any valuable subsystem of g much less of that of g as a whole[19]

Observe that in dictatorial political systems or respective societies something might be illegitimate that is legitimate according to D14. For example practicing religion (without fundamentalist or terroristic activities) is legitimate according to D14 but might be either forbidden by law or persecuted although permitted (legitimate) in such political systems.

Apart from these general higher values there are also personal values which play an important role for the higher level of QL.

DEFINITION 16 (D17). v is a *personal value* for person x who belongs to society g iff

1. v is desired by x or is thought to be desirable by x

2. v is legitimate in g

3. v can be realised by x to the best of x's knowledge and belief

Examples: choice of a profession, choosing a mate, choosing studies ... etc.

In addition there are religious values which are important for the quality of life of religious people.

DEFINITION 17 (D18). v is a *religious value* for person x who belongs to society g iff

1. =1. of D17

2. =2. of D17

3. v can be realised by x according to x's religious belief (faith)

Examples: life after death, life with love and joy without pain, poetic justice, worship of god ... etc.

According to the above definitions we can say what we understand by a higher level of quality of life:

---

[19]The definition of "legitimate" is due to Bunge (1989) p. 35.

DEFINITION 18 (D19). *hl* is a higher level of quality of life for person *x* iff *hl* is the realisation of

1. a higher value for *x* (D15) or

2. a personal value for *x* (D17) or

3. a religious value for *x* (D18)

4.4 Projected action space and projected evaluation

The effects of a person on its environment have been described by D6 (active action space) as effects factually occurring. But there are also *assumed* or *believed* effects: The respective person thinks, assumes, believes, is of the opinion that his (her) action has a certain effect on his (her) environment. This can in fact be true or can be only partially true or can be an illusion. An example would be if a politician thinks that his talk had a great effect on his audience concerning his political desires. The action space which is produced by such beliefs or opinions can be called the projected action space.

DEFINITION 19 (D20). Let x be a human person. Then the projected action space of x are those things y ($x \neq y$) such that:

x believes that x or a part of x acts effectively on y.

Observe that y need not necessarily belong to the environment of x (according to definition D1) because the effects of x on y need not be real effects, since they are believed effects and can be real, partially real or illusionary.

Examples: The effects of the speech of the politician, the advice of the educator, the order of the parents, the impression of the applicant ... etc. according to the opinion of the politician, educator, parent, applicant.

The projected action space can also be connected to living space 2 in the sense that the believed effects of an action contribute positively or negatively to the quality of life. This already shows that also evaluations are connected with the projected action space. There are several interesting examples where both means - end relations and values are projected on a value-neutral situation. The situation is value-neutral if it is viewed only from a factual point of view. To make this point clear we may refer to Braitenberg's book "Vehicles".[20] One of the experiments described in the book are the movements of a triangle, a circle

---

[20]Braitenberg (1984)

and a rectangular on a plane surface. Several of these movements are immediately interpreted as purposeful and to the (geometrical) figures good or bad intentions are attributed. This is the case, for example, if the triangle approaches the circle with one of its sharp corners, but the rectangle steps in between.

Concerning projected values Boesch described[21] at length the phenomenon of reading subjective valuations into one's environment or rather (more restrictedly) into one's living space. He distinguishes there objective components of meaning (which he calls denotations) from subjective ones (which he calls connotations): "However, all these denotations are associated with personal experiences, personal or cultural beliefs and evaluations. Thus the home forms a base for planning and initiating action, as well as a target for terminating it. These home-valences, obviously, are related to external valences of places, objects and people ... In other words, the things surrounding us will be increasingly structured into patterns; the home not only forms a structure of intimate familiarity, but by its power to exclude non-desired or threatening impacts of the external world takes on the quality of shelter."[22]

The mentioned projected evaluation can be characterised in the following way:

DEFINITION 20 (D21). Let x be a human person. Then the projected evaluation of x are those things y ($x \neq y$) such that:

1. y belongs to the projected action space of x

2. x interprets y as a certain value for x or x subordinates y under a hierarchy of values for x

4.5 Well-Being

According to the usual understanding it seems reasonable that *well-being* requires more than the satisfaction of the basic values (D11), i.e. more than reaching the basic level of quality of life (D12). In an analogous way, *well-being* requires more than the realisation of mental basic values (D13) or a higher level than the basic level of mental quality of life (D14). We think that well-being needs the realisation of a higher level of quality of life (recall definitions D15-D19). Moreover it will not be sufficient that this higher level of quality of life rests only a short

---

[21] Boesch (1991)
[22] ibid. p. 23 and 33.

time. On the contrary, to speak of well-being this higher level of quality of life should be preserved for a longer period of time. Thus we require these two conditions for well-being:

1. the realisation of values for the respective person clearly exceeds the realisation of basic values and mental basic values

2. this realisation of higher values is preserved for a longer period of time.

5 Lack and loss of quality of life and well-being

We want to distinguish lack of quality of life on the one hand and loss of the other. We use "lack" if there is a serious or basic defect w.r.t. quality of life. And therefore we understand by "lack of quality of life" an absence of one or more basic values (cf. D11) or mental basic values (cf. D13). On the other hand we understand by "loss of quality of life" or "loss of well-being" an absence of one or more higher values that exceed the basic values (cf. D15-D19).

A further distinction is necessary here. Lack or loss can be understood in a twofold way: First as absence of some value that should be present in order to guarantee the respective level of quality of life. Secondly in such a way that the absence of some value is put up in order to achieve a higher value or a higher level of quality of life. An example for the second case is an operation. The loss of being uninjured is put up in order to regain a basic value (health). This kind of loss plays an important role in everyday life.[23]

These considerations lead to the following definitions:

DEFINITION 21 (D22). l is a lack of a basic level of QL for x (concerning basic values or mental basic values) iff

it is necessary for x to avoid l in order to reach or to preserve this basic level of QL

Examples: lack of clean air or water, insufficient ground food, insufficient medical care . . . etc.

DEFINITION 22 (D23). s is a loss of a certain level A of QL for x iff

it is necessary for x to avoid s in order to achieve or to preserve level A of QL

Examples: loss of higher education, of professional training, of medical treatment, of the possibility of freely choosing ones religion . . . etc.

---

[23]Cf. the chapter on "necessary evil" in Weingartner (2003) p. 28ff.

DEFINITION 23 (D24). l is a purposive lack of a basic level of QL for x iff the following conditions are satisfied:

1 l is a lack of a basic level of QL for x (D22)

2 l is a lack of a basic value of x, but l contributes to avoid a greater lack of another basic value for x

3a l is necessary to protect or to achieve a basic value for x or

3b l is necessary to achieve or preserve a higher value for x

Examples: operation, fasting for reasons of health.

Condition 3b also allows putting up with a lack of a basic value in order to achieve a higher value. For example if parents put up with hard work that becomes a health hazard in order to feed children or to pay for their education. Or if students work hard the whole day in order to pay and join the evening/night courses at the university (Brazil).

DEFINITION 24 (D25). s is a purposive loss of a certain level A of QL for x iff the following conditions are satisfied:

1 s is a loss of level A of QL for x

2 s contributes to avoiding a greater loss of level A or of another level

3a s is necessary to achieve or preserve a basic value or

3b s is necessary to achieve or preserve a higher value of level A or another level

Examples: A master builder accepts also less gifted apprentices in order to help them to reach the completion of their professional training; he puts up with bungling and botching and with ruined work pieces in order to achieve the higher value of learning from mistakes; a mother interrupts her study at the university in order not to endanger care and education for the children. An example for 3b is if a parent changes his beloved profession to another one where he (she) gets more money to feed his children.

A purposive lack or loss is legitimate if it satisfies both conditions of definition D16. As has been mentioned these conditions can be violated in dictatorial regimes such that for example the liquidation of elderly, ill and disabled people is interpreted as purposive lack of QL. Is there legitimate loss of QL which is not purposive? This is such a legitimate loss which satisfies condition 1. but neither 2. or 3. of D25. In fact there are such cases: For example a legitimate house-building on a place such that it destroys the view of the neighbour. Usually to build at that place will not be necessary to achieve a higher level of QL.

As there is loss of QL, there is also *loss of well-being*. This is the case if the realisation of values and goals for some person does not exceed essentially above the basic values or the basic level of QL. Similarly to D25 we can understand *purposive loss of well-being*.

It is worth mentioning moreover that there is a kind of loss of well-being such that the respective higher values are no more desired. This can happen even with basic values in extreme cases. Such a loss can develop as a gradual weakening with doubts about values.[24]

Acknowledgements

We are grateful to Prof. Peter Weichhart for valuable comments.

BIBLIOGRAPHY

[1] E. E. Boesch. *Symbolic Action Theory and Cultural Psychology*. Springer, 1991.
[2] V. Braitenberg. *Vehicles*. Cambridge, 1984.
[3] M. Bunge. *A World of Systems. Treatise on Basic Philosophy Vol. 4*. Reidel, 1979.
[4] M. Bunge. *Ethics - The Good and the Right. Treatise on Basic Philosophy Vol. 8.*. Reidel, 1989.
[5] R. E. Dunlap and W. R. Catton. Which Function(s) of the Environment Do We Study? A Comparison of Environmental and Natural Resource Sociology. In M. R. Redcliff, G. Woodgate, editors, *New Developments in Environmental Sociology. Elgar Reference Collection*. Cheltenham, UK; Northhampton, MA, USA, 2005.
[6] V. E. Frankl. . . . *Saying Yes to Life in Spite of Everything: A Psychologist Experiences the Concentration Camp*. 2005.
[7] V. E. Frankl. *On the Theory and Therapy of Mental Disorders. An Introduction to Logotherapy and Existential Analysis*. Translated by James M. DuBois. Routledge, 2004.
[8] K. Friedrichs. Umwelt als Stufenbegriff und als Wirklichkeit. *Studium Generale*, **3**: 70-74, 1950.
[9] P. Handke. *A Sorrow Beyond Dreams*. NYRB Classics, 1972.
[10] K. Lewin, K. *Field Theory in Social Science*. Harper, 1951.
[11] S. Lofti. The 'Purposiveness' of Life. *The Monist*, **93**(1):123-134, 2010.
[12] M. Mahner-Bunge. *Foundations of Biophilosophy*. Springer, 1997.
[13] T. Oakley. The Issue of Meaninglessness. *The Monist*, **93**(1):106-122, 2010.
[14] K. R. Popper and J. C. Eccles. *The Self and its Brain*. Springer, 1977.
[15] J. v. Uexküll. *Umwelt und Innenwelt der Tiere*. Berlin, 1909.
[16] P. Weichhart. Remarks on the Term "Environment". *GeoJournal*, **3.6**:523-531, 1979.
[17] P. Weingartner. *Logisch-Philosophische Untersuchungen zu Werten und Normen*. Peter Lang, 1996.
[18] P. Weingartner. *Evil. Different Kinds of Evil in the Light of a Modern Theodicy*. Peter Lang, 2003.

---

[24]In his novel, Handke (1972) describes such a case. See also Oakley (2010).

Circumveiloped by Obscuritads:
The nature of interpretation in quantum mechanics, hermeneutic circles and physical reality, with cameos of James Joyce and Jacques Derrida[1]

F.A. MULLER

ABSTRACT. The quest for finding the *right* interpretation of Quantum Mechanics (QM) is as old als QM and still has not ended, and may never end. The question *what an interpretation of QM is* has hardly ever been raised explicitly, let alone answered. We raise it and answer it. Then the quest for the right interpretation can continue *self-consciously*, for we then know *exactly* what we are after. We present a list of minimal requirements that something has to meet in order to qualify as *an interpretation of QM*. We also raise, as a side issue, the question how the discourse on the interpretation of QM relates to hermeneutics in Continental Philosophy.

1 Nuemaid Motts and a Nichtian Glossary

James Augustine Aloysius Joyce (1882–1941) constructed this tantaltuous and tumulising towertome *Finnegans Wake* (1939) in the period during which quantum mechanics was *created* (1923–1939), by Werner Heisenberg, Max Born, Pascual Jordan, Wolfgang Pauli, P.A.M. Dirac and Erwin Schrödinger, was *axiomatised*, by Johnny von Neumann, was *applied*, by numerous physicists, was *interpreted*, by Niels Bohr and Heisenberg, was *demonstrated to exclude* certain alternative theories, by Von Neumann, and was *criticised*, by Albert Einstein, Schrödinger and others. All of this has continued and is continuing until the present day, including ever more and more physicist. Over the past decades,

---

[1] The main title of this paper is borrowed from James Joyce's *Finnegans Wake* (1939), 244.15), as are the titles of the Sections, for reasons that will become evident to the imaginative mind as we proceed; the notation '244.15' is standard and means: page 244, line 15. Any edition can be consulted, because they all use the same pagination and lining.

philosophers have joined the interpretation effort — with remarkable success, we dare add.

News broadcast in *Finnegans Wake*:[1]

> The abnihilization of the etym by the grisning of the grosning of the grinder of the grunder by the first lord of Hurteford expolodonates through Parsuralia with an ivanmorinthorrorumble fragoromboassity amidwhiches general uttermost confusion are perceivable moletons scaping with mulicules ... Similar scenatas are projectilised from Hullulullu, Bawlawayo, empyreal Raum and mordern Atems.

Recall that in 1911 Lord Rutherford (lord of Hurteford) split the atom (etym), a detonation of sorts where electrons (moletons) and molecules (mulicules) escape, projectiles moving through empirical space (German *Raum*). The historical event was reported all around the globe, like in Paris (Hullulullu), Rome (Bawlawayo) and Athens (Atems).

Traces of both the Quantum and the Relativity Revolution in physics are scattered all over *Finnegans Wake*.

Philosophically, *Finnegans Wake* can be seen to raise the issue of what *meaning* is, even of what *language* is. We do not awaken this grand defining issue of the philosophy of language.

To interpret a word, an expression, a sentence, a text, is *to assign meaning* to it. Clear and unambiguous kinds of texts, such as the telephone directory of Hullulullu, the weather forecast for tomorrow in Bawlawayo, the papers of Patrick Colonel Suppes, and the user manual of your brand new ten-dimensional retina-screen nanowave stringphone, do not stand in need of interpretation. Other kinds of text cry out earsplittingly for interpretation, of which *Finnegans Wake* arguably is the most clear and unambiguous instance ever created. An exposition of quantum mechanics is, like novels and poems, somewhere between text that need no interpretation and texts that absolutely require interpretation, although admittedly it will be closer to the aforementioned than to the last mentioned. As Dummett (1925–2011) testified:[2]

> Physicists know how to use quantum mechanics and, impressed by its success, think it is *true*; but their endless debates about the interpretation of quantum mechanics show that they do not know what it *means*.

But standing in need of interpretation is something that *Finnegans Wake* and quantum mechanics (QM) share with lots of other texts, such as the earlier mentioned novels and poems. We must take a closer look

---

[1] Joyce [1938], 353.22–23.
[2] Dummett [1991], p. 13.

to understand why they are special, to seek out what it is specifically that they have in common besides requiring (much or some) interpretation.

Joyce judged *natural-language-as-we-know-it* (whenceforth: Nalasweknowit) inadequate to describe what happens in the dream world, and created, for this very purpose, a 'new language', *if* that is the appropriate phrase, given how Joyce characterised his means of expression in *Finnegans Wake*: "nuemaid motts truly plural and plusible" (138.08–09) and a "nichtian glossery which purveys aprioric roots for aposteriorious tongues this is nat language in any sinse of the world" (83.10–11). Heisenberg and Bohr judged Nalasweknowit, of which they considered 'the language of classical physics' a refinement, inadequate to describe what happens in the microphysical world, the world of *very* small physical entities and *very* brief physical processes.

Small wonder. Nalasweknowit has developed while *homo sapiens* and its ancestry was wide awake, i.e. not dreaming, and interacting with the macrophysical world filled with trees, rocks and animals, and with days, seasons and lifetimes. Man was occupied with fulfilling his biological needs of nutrition, protection and procreation, which we share with beasts, rather than with penetrating the ephemerally flashing realm of dreams, explaining the phenomena by means of theories, or unravelling the mysteries of a realm of reality inaccessible by the unaided senses. No one had ever wanted or needed to go above and beyond the waking macrophysical world, or to transcend our biological needs. But, at some day, the time had come that we did want and did need to go precisely there, and we did want to transcend our beasty needs. How did we do it?

Back to the early 20th Century. Understanding the microphysical world was no longer deemed possible with Nalasweknowit. In order to grasp this realm of reality somehow, only a 'symbolic description', or a *Deutung*, by abstract mathematical means seemed possible. Of course nothing remotely like "multimathematical immaterialities" (394.31–32) were the means for Joyce to penetrate the realm of dreams. Joyce constructed numerous neologisms and *portmanteaux*: "the dialytically separated elements of precedent decomposition for the verypetpurpose of subsequent recombination" (614.34–35). In contradistinction to how Joyce accomplished his daunting task of evoking the phantasmagorical events of deep weep sleep, i.e. by creating *Finnegans Wake*, and thereby replacing Nalasweknowit, what the founding fathers of QM did was something far less radical: a comparatively small yet significant enrichment of Nalasweknowit would initially turn out to be sufficient to unlock the secrets of the atom and to enter the suprasensical world, —

but would eventually and entirely unexpectedly also ushered to perplexities the world of physics had never seen before ...

2 Abnihilazation and Everintermutuoemergent

No matter how one characterises QM pecisely, e.g. as the deductive closure of a set of sentences (the postulates) in a formal language or through a class of models (structures in the domain of discourse of axiomatic set-theory), or some sophisticated combination of these, or a some category with objects and arrows (to show the world proudly you're in full command of the latest thing), QM incontestably has *propositional content*, expressed in declarative sentences of Nalasweknowit, enriched with physical and mathematical vocabulary, symbols included. QM makes a large variety of pronouncements about physical reality, measurements included, that can be and have been tested severely. Sometimes QM says things that raise our eyebrows sky high, like there be non-local correlations that do *not* fall off with distance and *cannot be explained* even by an appeal to the entire past of the carriers of the correlata (version of Bell's Theorem), and like a continuously *observed* kettle filled with water on the fire that never boils (quantum Zeno paradox).[3] Sometimes QM remains mute when we desperately crave for answers, like when we ask whether Schrödinger's unmeasured, and therefore unobserved cat *is* dead or alive, since QM does neither fulfil the truth-condition for the sentence 'The *unobserved* cat is alive', nor for 'The *unobserved* cat is dead'. Here QM falls silent. Needless to add that the celebrated case of Schrödinger's cat extrapolates to the entire unmeasured part of the universe, which comprises nearly everything. We observe a few drops of the ocean of being. Nearly all of physical reality is *ontically indeterminate*. Therefore to speak of 'the measurement problem' is peculiar, and an historical accident in fact. The problem is better be called *the reality problem* of QM. QM forbids us to speak whereof we want to speak: reality.

Notice parenthetically that a use theory of meaning, which takes the use of words, expressions and sentences constitutive for their meaning, does not sit comfortably with Dummett's locution displayed above: if, *first*, knowing the meaning of QM resides in knowing how to use it, and, *secondly*, granted that physicists know how to use QM in every which way, that is, knowing how to construct quantum-mechanical models of phenomena, knowing how to reason quantum-mechanically, knowing

---

[3] For the sake of clarity: we suppose that to observe is to measure, so by contraposition, not to measure is not to observe; to measure is not necessarily to observe. This is correct, because think of, say, measuring the presence of a neutrino or the energy of an electron, which are unobservable entities: we measure but cannot observe.

how to calculate measurement outcomes, and knowing how analyse experiments using QM, then they should *know* its meaning, whereas the endless debates about the interpretation of QM — which we shall provisionally call its *hermeneutic predicament* — is taken to show the contrary, namely that they *do not know* what QM means.

If the project *to interpret* QM is, in good hermeneutic fashion, to assign meaning to it, we must ask which expressions of QM stand in need of interpretation, because, then, apparently *their* meaning is not obvious, or is ambiguous, or is obscure, or in any way stands in dire need of receiving clear and unambiguous meaning. If every expression in QM were perfectly clear, there would obviously be no need to interpret QM and there would not have been an interpretation debate. But there is.

QM is always presented in natural language enhanced by scientific vocabulary, mathematical as well as physical, many items expressed by symbols. Logical concepts are also be prominently present: and, or, not, there is, for all, implies, follows from, is consistent with, if–then, etc. So the vocabulary of QM can be subdivided in:

(α) Logical Vocabulary,
(β) Mathematical Vocabulary,
(γ) Physical Vocabulary.

We shall take these in turn.

(α) The default option about the Logical Vocabulary that standard classical (sentential and predicate) logic can be used to make it precise. But this is already not entirely uncontroversial: some interpretations of QM require deviations from classical logic, e.g. quantum logic, others do not. We shall proceed with the default option, and addres below the issue whether 'a change of logic' counts as an interpretation of QM. Further, it is well-known that *scientific inference* is not exhausted by deductive reasoning: abductive, inductive, probabilistic and analogical reasoning are abound in science. To capture these rigorously is an ongoing effort in philosophy of science. Since this topic has little if anything to do with the interpretation of QM, we gloss over it.

(β) The words in the Mathematical Vocabulary are crystal clear. They do not stand in need of interpretation — Hilbert-space, self-adjoint operator, eigenvalue equation, unitary evolution, statistical operator, Clebsch-Gordan coefficients, Weyl rays, unitary representations of a symmetry group, permutation operators, Wigner distributions, complex square-integrable function of n real variables, Von Neumann rings of operators, canonical commutator, and what have you.

(γ) The Physical Vocabulary includes: physical magnitude, physical system, composite system and subsystem, physical property and physical relation. These concepts also seem far from obscure. This is not to say that these concepts are beyond interpretation, let alone beyond *metaphysical* analysis. But debates in metaphysics on the nature of properties, relations (and existence for that matter) have had no bearing on debates on the interpretation of QM. Perhaps they should have? There seems little to say in utter generality. Any interpretation of QM that specifies conditions for the ascription of properties to physical systems can have a *realist* reading, which considers properties as abstract entities that are instantiated by physical systems, as well as a *nominalist* reading, which considers the mentioned conditions as application conditions for predicates. Such differences yield different interpetations of QM, but these differences are due to a prior metaphysical view that is independent of, and likely cannot be motivated by an appeal to QM. Such differences in interpretation are uninteresting for those interested in QM. Two interpretation of QM who differences are all due to such prior metaphysical views are hereby distinguised and deserve a special name: we shall call them *ininteresting* variants of each other.[4]

Physical concepts of QM that stand in need of interpretation are: ($\gamma.i$) certainly the concept of measurement itself, ($\gamma.ii$) the probability for finding specific outcomes upon measurement, and ($\gamma.iii$) physical state.

($\gamma.i$) On the one hand, one can send everybody who raises questions about measurement to a laboratory: observe what is happening there and ask around; if that will not do, then nothing will. On the other hand, when we ask what a measurement is, we are after a general answer, a general concept of measurement, one that encompasses what happens inside all laboratories; everything we want to call a measurement should be an instance of our general concept, and everything we do not want to call a measurement should not be an instance of it. This general concept should cover our use of the word 'measurement', but need not cover it entirely, for we shall gladly pay the price of lack of full coverage for a clear and distinct general concept. In short, we are after a Carnapian explication of the concept of measurement. By way of an interjection, we shall attempt this in the next Section. When a physical system qualifies as piece of measurement apparatus, when a physical interaction qualifies as a measurement interaction, when an event qualifies as a measurement event, and perhaps more, have been issues for analysis and controversy since the advent of QM. Certainly we want to count these issues part and parcel of the discourse 'the in-

---

[4]Rather than 'uninteresting': I have replaced the suffix 'un' with 'in', alluding to 'interpretation'.

terpretation of QM'.

(*γ.ii*) Probability is mathematically represented by a normed additive mapping from some Boolean subset family of $\mathbb{R}$, say the intervals $\mathcal{I}(\mathbb{R})$, to the interval $[0,1] \subset \mathbb{R}$:

$$\text{Pr}: \mathcal{I}(\mathbb{R}) \to \mathbb{R}. \tag{1}$$

So for the mathematician, this is all there is to probability: a normed measure on $\mathcal{I}(\mathbb{R})$; probability is a branch of mathematical measure theory. Not for the scientist, who has to relate this normed measure to the world. The quantum-mechanic has to relate probability at the very least to measurement outcomes. The only way to do this is via relative frequencies. But whether probability *is* a limiting relative frequency amounts to taking a further philosophical step, as does identifying probability with *objective change*, as does identifying it with *propensity*, i.e. some *generalised quantitative disposition*, and as does taking it as a *degree of belief of a human being* or a *degree of rational credence*. We have now entered the field of the interpretation of probability. Some hold that *quantum probability* is somehow special and different from probability as it occurs elsewhere in physics and in science generally. The challange then is to explain wherein this difference resides.

(*γ.iii*) The physical state of a physical system can best be taken as a *primitive* concept, which can not be analysed further into other physical concepts. In QM, it can be and is represented mathematically in a variety of distinct ways: as a

- ⊛ a normed Hilbert-vector, or
- ⊛ a Weyl-ray, or
- ⊛ a statistical operator acting on a Hilbert-space, or
- ⊛ a positive map on a C^*-algebra, or
- ⊛ a set of probability measure.

Maybe calling a Hilbert-vector (or Weyl-ray, or ...) *the mathematical representative of the physical state* of a physical system is a *mistake*: a Hilbert-vector should remain a physically uninterpreted and purely mathematical concept in QM, an auxiliary device to calculate probability distributions of measurement outcomes. There is no 'physical state' of the unmeasured cat in purgatory: we are led to believe that the cat has, or is in, a *physical state* by mistakenly trying to attribute physical meaning to a Hilbert-vector that is a superposition of two vectors, which according to the standard property postulate we associate with a cat having the property of being dead and one having the property of being alive, respectively. We believe the unmeasured cat is some particular physical state but perhaps it isn't. QM then associates a Hilbert-vector to the cat that is devoid of physical meaning, but enables the computation of probability measures over measurement-outcomes, which are filled with physical meaning. Thus we have physical meaningfulness out of physical meaninglessness. Sheer magic. Sadly magic does not help us to understand physical reality.

The wilful jump to meaninglessness seems however a cheap way out. I don't like it. We believe that the unmeasured cat is either stone dead or breathing, because *tertium non possibilium*, and we want QM to be logically compatible with this belief, at the very least, and preferrably to imply one or the other belief.[5] After all, QM also predicts that as soon as we peek at (i.e. measure) the cat, through a pinhole, unbeknownst to the cat, it *is* either dead or alive. Rather than to withhold physical significance from the Hilbert-vector, we should try to assign physical significance to it (or to a Weyl-ray, or ...). For how else could it determine physically meaningful probability measures over measurement-outcomes? No physical significance in, but physical significance out? That ought to be unacceptable. One way is to connect Hilbert-vectors to equivalence classes of preparation procedures in the laboratory. This won't help us however with Schrödinger's unmeasured cat. This won't help us with anything, because superpositions are the rule, not the exception. The founding fathers of QM started with electrons in superpositions, soon other elementary particles followed, then atoms, and nowadays we have bucky-ball molecules and circulating currents in superconducting metals in superpositions in the laboratory. The march of superpositions from the realm of the tiny to the realm of medium-sized dry objects is not halting.

So-called *modal interpretations* of QM have taught us that the cat ceases to be a problem as soon as we reject 'half' of what we shall call the *Standard Property Postulate* of QM, which one can find the classic texts of Von Neumann [1932] and Dirac [1928] — and which remains nearly always tacit in textbooks on QM.[6]

■ **Standard Property Postulate (Dirac, Von Neumann).** *A physical system S having physical state $|\psi\rangle \in \mathcal{H}$ has quantitative physical property mathematically represented by the ordered pair $\langle B, b \rangle$, where B is an operator representing some physical magnitude and where $b \in \mathbb{R}$, iff $|\psi\rangle$ is an eigenstate of B having eigenvalue b: $B|\psi\rangle = b|\psi\rangle$.*[7]

---

[5]The border between dead and alive may be vague, in which case read for 'alive': not dead.

[6]Any author on QM who presents Schrödinger's cat as a problem in that it is neither dead nor alive, tacitly assumes that it is necessary for the cat to be in a relevant eigenstate in order to be either dead or alive. The Standard Property Postulate is also known as 'the eigenstate-eigenvalue link'.

When it is no longer necessary for the state to be an eigenstate of B in order for physical system S to have a property of the sort $\langle B, b \rangle$, then the unmeasured cat *can* be either dead or alive even when its state is *not* a corresponding eigenstate — but is a superposition of such eigenstates. The compatibility between QM and our belief that the unmeasured cat is either dead or alive is saved. What can be adhered to, then, is not the Standard Property Postulate but the ∎ **Sufficiency Property Postulate**, according to which it is sufficient (but not necessary) for the system to be in some eigenstate of B in order to possess property $\langle B, b \rangle$ (one drops one conjunct of the Standard Property Postulate).

Logically weakening a postulate seems however to have little to do with *interpretation* in the hermeneutic sense of assigning meaning to expressions whose meaning is unclear, ambiguous or obscure. Indeed, for modal interpreters of QM, the problem of interpretation is to find *the right conditions for property ascriptions* — in addition to the stingy Sufficiency Property Postulate —, rather than to dwell on the meaning of 'physical state' (we say 'stingy', because a physical system is almost never in an eigenstate, so one can almost never invoke the Sufficiency Property Postulate). This points away from hermeneutical activity when considering interpreting QM — unless one subscribes to a theory of meaning such that changing the conditions for the ascription of properties changes the meaning of the word 'property', in which case one should consider such property postulates as Carnapian *meaning postulates*, rather than synthetic postulates that are either made true or made false by the way the world is.

It is in order to mention the exception of Oxonian Everettians, who under the lead of S.W. Saunders tinker with the meaning of 'existence' and tensed expressions by relativising them to a 'perspective', a 'branch', and who, like all Everettians, assign special significance to the terms of the state vector when expanded in a special basis, which is selected by the physical proces of decoherence.[8] They re-interpret and therefore change the meaning of words in Nalasweknowit. Hermeneutics in action. One could also maintain that the problem of interpreting QM just is the problem of finding an intelligible physical meaning to attribute to the mathematical concept of a Hilbert-vector (or ...) in such a way that our belief that the unmeasured cat is either dead or alive survives whilst leaving the Von Neumann postulates of QM untouched in all their glory, save perhaps minor modifications. But then modal interpreters of QM are *not interpreting* QM. There is no hermeneutic activity going on. What, then, *are* they doing?

---

[8]See Wallace [2013] for a state of the art defence of the Everett Interpretation.

They are changing the theory of QM by *changing* (one of) its postulates, which results in a *different* theory of QM, just like changing the parallel axiom of Euclidean Geometry results in a *different* geometrical theory. When that different geometrical theory, if true, tells us that the structure of space is different from what Euclidean Geometry tells us, then *mutatis mutandis* modal QM provides a different description of the microphysical world than standard QM does. This is the key insight of this paper and the essence of our alternative view of what it means *to interpret* QM. But before we turn to that, first the promised interjection on $(\gamma.i)$ measurement.

3 Multimathematical Murkblankered Immaterialities

3.1 Preambule

In English, as in most languages, *to measure* is a *verb*. The *noun* 'measurement' is derived from it: to perform a measurement is synonymous with to measure. To measure is a manifestation of intentional behaviour, i.e. it is a type of *action*, performed by a human being, with a purpose — or by any being having the cognitive capacities to exhibit 'measurement behaviour'. Therefore the concept of measurement is an *intentional* concept.

The concept of measurement is expressed most explicitly, we submit, by a pentatic predicate: *someone* (p) measures *something* ($\mathcal{A}$) that pertains to *something* (S) using *something* else (M) and obtains *result a*:

$$\text{Measure}(p, \mathcal{A}, S, M, a) : \; p \text{ measures } \mathcal{A} \text{ of S by means of M and obtains } a. \tag{2}$$

There are *kinds* of measurements, whose extensions are subclasses of the extension of (3.1): demolishion measurements, ideal measurements, extensive measurements, perfect measurements, sharp measurements, weak measurements, ... The word measurement occurs in combinations with other words, especially in science; these combinations express different but allied concepts, which we call *measurement concepts*: measurement event, measurement process, measurement procedure, measurement result, measurement outcome, measurement interaction, measurement apparatus, measurement theory, measurability. In every case, the suffix 'measurement' points to a *kind*: measurement events are a *kind* of events, they form a subclass of the class of all events; measurement processes are a *kind* of processes, they form a subclass of the class of all processes; *etc*. The purpose of this Section is to analyse the concept of measurement (3.1) the core concept Meas (3.1). The other measurement concepts will have to wait (they can however easily be defined in terms of Meas).

In our *analysandum*, the concept of Measurement (3.1), five things are connected: human being p, value a, entity S, entity M, and magnitude $\mathcal{A}$. The challenge is to characterise these concepts in a way that does not rely on any measurement concept, otherwise we awaken the spectre of circularity. We gloss over the concept of a human being and move now to the other concepts from the putative *analysans*, one per Subsection.

3.2 Values

Value a is a number. Number a is a rational number ($a \in \mathbb{Q}$), because every measurement has a finite accuracy, i.e. a finite number of significant digits. Since two measurement results, a and b, can be taken as the real and the imaginary part of a complex number, there is room for extending $\mathbb{Q}$ to $\mathbb{C}_{\text{rat}} \subset \mathbb{C}$, the set of complex numbers having rational real and imaginary parts. Nonetheless we continue with $a \in \mathbb{Q}$ and bracket $\mathbb{C}_{\text{rat}}$.

To count is also a form of measurement, with a natural number as the result. One can count the number of children in the class room with infinite accuracy: there are 23 children in the class room, or $23,000\ldots$, and not 23 ± 1, let alone $23,0 \pm 0,2$. (In these cases, the outcome still is a rational number, because $\mathbb{N} \subset \mathbb{Q}$; so we can stick to $a \in \mathbb{Q}$.)

3.3 Entities

We measure the emission spectrum *of* Hydrogen; we measure the mass *of* the Earth or *of* a positron; we measure the intensity *of* the radioactive radiation *of* the nuclear power plant in Harrisburg; we measure the acidity *of* the liquid in this flask; *etc*. Clearly *what* we measure, $\mathcal{A}$, always pertains to something (S), and that something, that entity, we take to be a *physical system*, as broadly construed as possible: it consists of matter and fields, and is located in space-time. This makes physical systems, in metaphysical parlance, concrete rather than abstract entities.

3.4 Measurement Apparatus

A measurement apparatus also is a physical system, that much seems clear. We thus need a criterion to tell us *which* physical systems qualify as a measurement apparatus. We proceed stepwise, **(i)–(iii)**: in each step we consider a concept that we shall use in characterising what a measurement apparatus is.

(i) *Observability.* Surely a measurement apparatus M is a physical system that we, human beings that measure, should be able to see (or hear ...). Otherwise M is of no use to us! So M has to be *observable* by us. This raises immediately the further question which physical systems are *observable*. Philosophers of science have pondered this question. We shall mention here the rather obvious philosophical criterion for the extrinsic property of observability. Let p be a normal person, of sound mind and having normal eye-sight.

> **Criterion for Observability.** Physical system S is *observable* iff for every p: if p were in front of S in broad daylight with open eyes and looks at S, then p would see S.

Van Fraassen famously insisted that the observability of objects, events, facts, processes, is a subject for scientific research, not for philosophical analysis. For a scientific characterisation of observability, see Muller [2005].

If S is observable, then S seems to have *properties* that are observable, notably its shape and colours. What *is* it that we actually *see*: the object or its properties, or both? In full generality, this is a metaphysical question, which we wish to bracket. We therefore limit ourselves to a characterisation of an observation predicate, remaining neutral about whether predicates express universals or tropes.

> **Criterion for an Observation Predicate.** A predicate F applied to observable physical system S is an *observation predicate* iff for every person p: if p were in front of S in broad daylight with open eyes, then p would immediately judge that $F(S)$ or immediately judge that $\neg F(S)$ relying only on looking at S and not in addition on some theory; such judgements we call *observation judgements*.

The addition of 'and not inaddition on some theory' is to prevent that *theory*, broadly construed, is relied on in order to judge whether $F(S)$ or that $\neg F(S)$. Consider a a Stern-Gerlach experiment, judging that 'the electrons have spin up' is not an observation judgement, because 'spin' is a predicate that cannot be understood without understanding some QM, and because electrons are unobservable, whereas judging that 'there is a black dot in the upper half of the white screen' is an observation judgement, with 'having a block dot in the upper half' being the observation predicate applied to 'the white screen', which is an observable physical system. We point out that person p must posssses a language, otherwise p could not form the judgement that $F(S)$ or that $\neg F(S)$. As soon as knowledge of a theory is needed to understand what predicate F means, F cannot be an observation predicate. Every observation judgement trivially is always 'concept-laden', because a predicate expresses a concept and an observation predicate is no exception. Judgements like 'the electrons have spin up', made while looking at the screen, is a 'theory-laden' judgement.[9]

So much for the observability of measurement apparatus M.

(ii) *One-one Correspondence*. When we read that the pointer of an Volt-meter points to 22 V, we ascribe the property of an electric potential difference to a circuit; when I read 86 kg on the display of a scale while standing on it, I conclude that my body has a mass of 86 kg; *etc*. So what we need is a one-one correspondence between observable properties of M and values of the magnitude $\mathcal{A}$ that M is measuring. Or better, *intervals of values* rather than single values because of the finite measurement accuracy: result $I = 1.04 \pm 0.07$ mA describes an observable property of an Am-meter that corresponds to an infinite set of electric current values, namely interval $[0.97, 1.11] \subset \mathbb{Q}$.

(iii) *Relevant Interaction*. So a measurement apparatus M of magnitude $\mathcal{A}$ is an observable physical system that leads to a one-one correspondence between certain sets of values of $\mathcal{A}$ and observable properties of M?

Almost right. McGuffey can assign a rational number to the three billiard balls lying on the table in front of him using pencil and paper: McGuffey looks at a ball and writes down some arbitrary rational number. McGuffey claims to have measured the masses of these balls, for we have a one-one correspondence between observable properties of the paper (the ink spots on it that express rational numbers) and values of the physical magnitude mass of the balls. Yet surely the pencil and paper do not qualify as a piece of *mass-measuring equipment*. Pencil and paper can be used *to report* measurement outcomes, but they are not themselves pieces of mass-measurement equipment. Furthermore, just writing down an arbitrary rational number with a pencil on a piece of paper is not measuring anything. McGuffey has *not* measured the mass of the billiard balls. If a one-one correspondence were enough, then measurement results would be what we want and choose them to be, and would become wholly under our control, whereas a measurement outcome seems to be something that is entirely beyond our control, something that has nothing to do with what we want. Particular measurement outcomes may be the ones we want, hope, wish, expect or fear. But *which* outcomes we shall actually obtain when we measure is beyond our control and indifferent to our needs, hopes, wishes, expectations and fears. Reality has a decisive say in it.

Perhaps we should require that the one-one correspondence *must be*

---

[9]Granted that it is controversial whether obseration predicates can be cleanly characterised and discerned from 'theoretical' predicates, we take sides in this controversy, in favour of there being clear and distinct observation predicates. For a recent defence, see Votsis [2015].

the result of a particular physical interaction between measured object S and measuring object M. McGuffey's one-one correspondence was not due to an interaction between the objects on his table and the paper. Which particular physical interaction? The physical interaction that occurs in explaining how M works, specifically how the one-one correspondence between (sets of) values of $\mathcal{A}$ and (observation) predicates that apply to M comes about. Let us call that physical interaction $\mathcal{A}$-*relevant* — which thus partly is an epistemic concept.

We arrive at the following criteria.

Criterion for an $\mathcal{A}$-Measurement Apparatus. Physical system M is an $\mathcal{A}$-*measurement apparatus* iff
(M1) M is observable; and
(M2) there is a one-one correspondence between (observation) predicates F which apply to M, and sets of values of $\mathcal{A}$; and
(M3) the correspondence of (M2) is the result of the $\mathcal{A}$-relevant physical interaction between physical system S, to which $\mathcal{A}$ pertains, and M.

Criterion for a Measurement Apparatus. Physical system M is a *measurement apparatus* iff there is some physical magnitude $\mathcal{A}$ such that M is an $\mathcal{A}$-measurement apparatus.

The young tree in the park garden is a measurement apparatus of the dichotomic physical magnitude 'presence of wind' ($\mathcal{W}$): if it oscillates visibly, then $\mathcal{W}$ has value 1 (presence of wind), and if it remains unmoved, then $\mathcal{W}$ has value 0 (absence of wind). Conclusion: a piece of measurement apparatus need not be a *technological artifact*, designed and constructed by human beings. Mother Nature produces pieces of measurement apparatus too, unintendedly, which is why being a technological artifact for M is not part of the criterion for a measurement apparatus.

3.5 Magnitudes

Etymologically the word 'magnitude' comes from the Latin *magnus* (big, large) and *magnitudo* (measure of bigness). Here 'measure' means *unit*, which suggests that magnitude is a quantified conception of some property: we speak of magnitude when we can quantify some property and we can measure it, no matter how indirectly. Think here of mass as quantity of matter (Newton), momentum as quantity of motion (Huygens), volume as quantity of 3-dimensional space, acidity as quantity of acid in a solution (Arrhenius), biomass as quantity of matter produced in carbon, hydrogen and oxygen, electric current as quantity of electricity (Gilbert), and so forth.

A general definition of magnitude is not around. An appealing idea seems to define a magnitude as a *quantified* or *quantitative property*. Measuring magnitude A of physical system S and obtaining value a would then show that S possesses a quantified property that we could represent by: $\langle A, a \rangle$. But this runs afoul against standard QM, which has taught us that measuring A definitely is *not* revealing a property possessed by S before the measurement. On the contrary, property $\langle A, a \rangle$ gets ascribed to S *just after* a measurement has ended and the state of S collapses to an eigenstate that belongs to value a, which then is an eigenvalue of the A-representing operator $\widehat{A}$ acting on the Hilbert-space $\mathcal{H}$ associated with S.

Thus we take magnitude A as primitive and define a *quantitative property* as $\langle A, a \rangle$, where $a \in V(A) \subseteq \mathbb{R}$, the set of values of A, or as $\langle A, a, u \rangle$ when magnitude A has a *unit* u. If needed, $V(A)$ can include complex numbers, in which case $V(A) \subseteq \mathbb{C}$.

A few examples ($\mathbb{R}^+$ contains 0):

$$\langle \text{mass}, \mathbb{R}^+, \text{kilogram} \rangle, \quad \langle \text{length}, \mathbb{R}^+, \text{meter} \rangle, \quad \langle \text{energy}, \mathbb{R}, \text{joule} \rangle. \tag{3}$$

We have now taken care of everything that is involved in the concept of measurement (3.1). Next we present our explication of measurement.

3.6 Main Dish

Much of the labour we had to perform to arrive at an *analysans* of our *analysandum*, that is, at a criterion for the core concept of measurement, has already been performed in our analysis of a measurement apparatus.

> **Criterion for Measurement.** *p measures A of S by means of M and obtains a* iff
> (1) p is a person,
> (2) A is a physical magnitude,
> (3) S is a physical system,
> (4) M is an A-measurement apparatus,
> (5) $a \in V(A)$, the set of possible values of A, and
> (6) p makes S and M physically interact A-relevantly, and this A-relevant interaction
>   results in A having value a, which M registers or displays.

Does this criterion cover all measurements that ever have been, are and will be performed, by anyone anywhere? I would be surprised if it did. For example, how about measuring the length of the table by a tapeline? Is the result of 250 cm (the value of the length of the table), a result of a 'lenght-relevant physical interaction between table and tapeline'? Their interaction consists of no more than they absorb some of each other's emitted electro-magnetic radiation... For another example, how about measuring time by a clock? When the clock is the measurement apparatus M, what is the physical system S? Perhaps also M: it measures the length of its worldline of spacetime, although that presupposes the Theory of Relativity. What we can do in the face of rods and clocks is to call the explicated concept above *Interaction Measurement*; measuring lengts and times by means of rods and clocks, respectively, the fall under the concept of Non-Interaction Measurement, a concept to be explicated in the future. But let's stop, and ask what a 'measurement interaction' is.

> **Criterion for Measurement Interaction.** A physical interaction I between two physical systems is a *measurement interaction* iff there is a physical magnitude A such that at least one of the physical systems is an A-measurement apparatus and I is the A-relevant physical interaction.

This characterisation of measurement interaction is not entirely physico-ontological but partly empistemological, just as measurement is, due to our characterisation of what an A-relevant interaction is (see above). This is how it ought to be, for to measure is to acquire knowledge. Measurement also counts as a species of knowledge acquisition. Quantum-mechanical measurement theory provides more detailed mathematical representations of measurement interactions, but it leaves the conceptof measurement, remarkably, un-analysed.[10] Back to the interpretation of QM.

4 Building supra Building pon the Banks for the Livers by the Soangso

The Prime Directive of Physics is that numbers calculated by using a physical theory (or model or hypothesis or principle) should coincide with numbers measured that pertain to physical systems the theory is supposed to be about. Suppose there is a minimal set of postulates of QM in the sense that the Prime Directive is obeyed: the postulates are just enough to calculate measurement outcomes and their probability measures, and these outcomes match what is being measured. Call this:

---

[10]See Suppes [2001], pp. 63–73, for some general Measurement Theory; see Bush, Lahti and Mittelstaedt [1996] for physical measurement theory. The concept of a measurement apparatus is not analysed but taken for granted in both books, remarkably.

minimal QM (QM$_0$, soon to be characterised rigorously).

The epistemic aim of physics can be several things:

- to save the (observed and unobserved) phenomena;
- to explain the (observed and unobserved) phenomena;
- to understand why things happen when and where they happen;
- to find out what physical reality is like, what it is made of, what there is, what exists, what the properties and relations are of the actual beings, and how the actual beings behave and influence each other;
- to reveal the structure of the universe as it is in and of itself;
- any other aim that goes above and beyond merely calculating putative measurement outcomes.

Save the first one, QM$_0$ falls short of reaching any of these aims of physics, for instance by telling us nothing about the fate of the cat and any other physical system that is not measured. QM$_0$ leaves too many meaningful questions about physical reality wide open. When QM$_0$ is a failure, must it not be refused entrance to the body of scientific knowledge? Is the current presence of QM$_0$ in that body not a cyst which should be surgically removed?

Nay nay, do not be afraid, I am not going to propose *that*. On the contrary, QM$_0$ will be the basis of it all.

Definitions. An *interpretation of* QM is another theory that is physically equivalent to QM$_0$, changes the vocabulary of QM$_0$, and extends the postulates of QM$_0$.

A theory is *physically equivalent to* QM$_0$ iff that theory is empirically equivalent to QM$_0$ and does not extend the physical vocabulary but may change the mathematical and logical vocabulary of QM$_0$.

A theory is *empirically equivalent to* QM$_0$ iff that theory is confirmed by exactly the same actual and possible measurement outcomes as QM$_0$.

An interpretation of QM must provide answers to questions about physical reality that we deem meaningful and that pertain to physical systems falling within the purview QM$_0$. We point out that our definition of an interpretation of QM is in harmony with what Van Fraassen says about it:

Suppose we agree that there can, in principle, be more than one adequate interpretation of a theory. Then it follows at once that interpretations go beyond the theory; the theory plus interpretation is *logically stronger* than the theory itself. For how could there be differences between views, all of which accept the theory, unless they vary in what they add to it?[11]

There may be hermeneutical activity in the wake of extending QM_0 in the literal sense of the word, in that the meaning of certain expressions has to be adjusted to fit the intended extension of QM_0, but the core interpretational activity is to extend QM_0 by adding postulates and vocabulary.

What is QM_0 precisely? Here follows an attempt to characterise it.

P0. Hilbert-Space Postulate (Von Neumann). *Associate some Hilbert-space $\mathcal{H}$ to physical system S, and a direct-product Hilbert-space to a composite physical system with the factor Hilbert-spaces being associated to the disjoint subsystems.*

P1. Evolution Postulate (Schrödinger). *Time is represented by the real continuum ($\mathbb{R}$).*
*IF no measurements are performed in time-interval $\Delta \in \mathcal{I}(\mathbb{R})$ on physical system S, where $\mathcal{I}(R)$ is a Boolean subset algebra of closed intervals of $\mathbb{R}$, THEN at every moment in time $t \in \Delta$, associate a Hilbert-vector $|\psi(t)\rangle \in \mathcal{H}$ (**P0**) to S such that there is a connected Lie-group of unitary operators acting in $\mathcal{H}$ such that $|\psi(t)\rangle = U(t)|\psi(0)\rangle$, where $|\psi(0)\rangle$ is associated to S at time $t = 0$, and where $U(t)$ is a group member, obeying the equation: $U(t + t') = U(t)U(t')$, for every $t, t' \in \Delta$.*

P2. Magnitude Postulate (Von Neumann). *Represent physical magnitudes of interest by operators acting in $\mathcal{H}$ (**P0**) that have some spectral resolution. Restrict the domain of this resolution to $\mathcal{I}(\mathbb{R})$, so that we consider only: $\mathcal{I}(\mathbb{R}) \to \mathcal{P}(\mathcal{H})$, $\Delta \mapsto P^B(\Delta)$, where $P^B(\Delta)$ is a projector from the Hilbert-lattice $\mathcal{P}(\mathcal{H})$ that belongs to the spectral resolution of positive operator B.*

P3. Probability Postulate (Born). *The probability for finding a value in interval $\Delta \in \mathcal{I}(\mathbb{R})$, at time t, upon measuring physical magnitude represented by operator B (**P2**) when Hilbert-vector $|\psi(t)\rangle \in \mathcal{H}$ is associated to S at time $t \in \Delta$ (**P0**, **P1**), equals the expectation-value of $P^B(\Delta) \in \mathcal{P}(\mathcal{H})$; in Redhead-notation ($([B]^{|\psi(t)\rangle}$ is the value of B when S is in state $|\psi(t)\rangle$):*

$$\Pr([B]^{|\psi(t)\rangle} \in \Delta) = \langle \psi(t)|P^B(\Delta)|\psi(t)\rangle . \quad (4)$$

For the sake of brevity, we have left out the Symmetrisation Postulate, which is about composite systems of similar particles, although one should think of QM_0 as including it.

Notice that QM_0 only speaks of *physical systems, physical magnitudes and probability distributions over measurement outcomes*, which exhausts its physical vocabulary. The theory QM_0 is sufficiently strong to enjoy an extremely wide variety of confirmation. We point out that physical magnitudes can be identified with equivalence classes of measurement procedures, so 'physical magnitude' can be eliminated from the primitive physical vocabulary — alas at the price of adding 'measurement procedure'. Not a word in QM_0 about physical states, physical properties and physical relations. No not one. *Stricto sensu* QM_0 is a mathematical recipe to calculate probability distributions over measurement outcomes. QM_0 says little if anything about physical reality outside the laboratory, let alone about the microphysical world. This is unacceptable.

QM_0 does not include the notorious Projection Postulate (for a moderately precise statement, see below). Can QM_0, then, deal with repeated measurements and then obtaining the same outcome (for discrete spectra)? If not, QM_0 may be an empirical failure. An adherent of QM_0 may uphold that the relative frequencies of repeated measurements can be described by conditional probability measures. A critic may respond that conditionalising in QM is the same as applying the Projection Postulate. Etc.[12]

The game of physics. One group of people, the Experimenters, produce numbers by manipulating various technical artifacts, and another group of people, the Theoreticians, think of mathematical recipes that also produce numbers. The aim of the game is that those numbers should match. The Experimenters usually begin and the Theoreticians then must match whatever the Experimenters come up with. If the Theoreticians fail, they lose and the Experimenters win; if the Theoreticians succeed, they win and the Experimenters lose. Sometimes the Theoreticians begin and then the Experimenters have to match. This is the game of physics, even the game of science, in a nutshell I take it. But *why* do we play this game? Out of boredom? For the helluvit? I say: *No no no*. We play it because we want the Theoreticians to win, because when they win repeatedly with the same theory, that theory may be knowledge of physical reality, may provide explanations of the phonemena that make us understand physical reality, and gathering such knowledge is the epistemic aim of physics. Otherwise the repeated empirical success of the theory would be a miracle and we don't believe in miracles.

Standard, or *orthodox quantum mechanics* (QM_{ox}) qualifies as an inter-

---

[12]See e.g. Fraassen [1991], pp 227–231.

pretation in the sense above of being an extension of QM_0, for it enriches the vocabulary of QM_0, strengthens some of its postulates and adds new postulates to it.

The language of QM_{ox} adds: physical properties and physical state. The Hilbert-Space Postulate (**P0**) becomes the

■ **Pure State Postulate (Von Neumann).** *Every possible pure physical state of a physical system is mathematically represented by a normed vector in some Hilbert-space, which we associate with the physical system.*

(The 'pure' alludes to a more general State Postulate encompassing also mixed states, which are not mathematically represented by Hilbert-vectors. We gloss over this.) Also the Standard Property Postulate is added, as well as the controversial

■ **Projection Postulate (Dirac, Von Neumann).** *IF one performs a measurement of physical magnitude B on a physical system, when it has state $|\psi(t)\rangle \in \mathcal{H}$ at the moment $t \in \mathbb{R}$ of measurement, AND one finds outcome in $b \in \Delta \in \mathcal{I}(\mathbb{R})$, with Δ the measurement accuracy of measuring value $b \in \Delta$,*
THEN immediately after the measurement outcome $b \in \Delta$ has been obtained, the post-measurement state of the physical system is represented by $P^B(\Delta)|\psi(t)\rangle$.

The Probability Postulate (**P3**) entails that the probability of finding a measurement-outcome, when measuring physical magnitude B, that does not lie in the spectrum of B vanishes. Since it depends on one's interpretation of probability of whether it follows that finding a measurement-outcome that is not in the spectrum of B is impossible, an explicit postulate is needed to exclude this. Here it comes.

■ **Spectrum Postulate (Schrödinger, Von Neumann).** *All and only values from the spectrum of an operator that represents a physical magnitude are its possible measurement-outcomes.*

So much for minimal QM_0 and its standard interpretation (aka orthodox quantum mechanics: QM_{ox}). Let us turn for a moment to a few other interpretations.

5 The Ineluctible Morality of the Intelligible

Bohr's Copenhagen Interpretation has long been, and perhaps still is, the interpretation most physicists adhere to; it pervades textbooks on QM. It adds the following postulates to QM_0, resulting in, say, QM_{Cop}.

■ **Quantum Postulate (Bohr).** *Every quantum phenomenon is indivisible; disconnected considerations of its parts are inappropriate, because the interaction between object-system and preparation and registration apparatus is not eliminable due to Planck's constant* $(h > 0)$.

By the *quantum phenomenon*, Bohr means the whole of the preparation apparatus, which one uses to prepare the object-system in a particular physical state, the registration apparatus one uses to measure some physical magnitude, and of course the physical system that is being subjected to preparation and measurement, the *object-system*. In classical physics one can appropriately consider parts, without mentioning other parts or the whole. In QM_{Cop} the Quantum Postulate rules, which is however limited by the

■ **Buffer Postulate (Bohr).** *The literal description of preparation and registration apparatus, and of the measurement outcomes, is given in the language of classical physics; the* Deutung *of the object-systems proceeds by means of mathematical concepts.*

Finally there is the

■ **Complementarity Postulate (Bohr-Heisenberg).** *The quantum phenomenon, specifically the experimental arrangement of the pieces of measurement apparatus (preparation and registration apparatus), determines which classical concepts are applicable. There are pairs of classical concepts, like wave/particle, kinematics/dynamics, space-time/causality, that are never jointly applicable in a single experimental arrangement but only in mutually exclusive experimental arrangements and in this way provide an exhaustive description of the object-system. Such pairs are called* complementary. *They are however jointly applicable in so far as the relevant Indeterminacy Inequality permits.*

According to Bohr, the language of classical physics is indispensable for QM. Bohr viewed this language as a *refinement* of Nalasweknowit: material objects in space, that persist over time and whose properties change over time as a result of causal processes. The 'classical language' is unambiguous and accurate, so that the objectivity of QM is guaranteed.

By *classical science* in general, Bohr meant scientific inquiry where the role of the scientist, the subject and his thought and talk, can be ignored, thus resulting in a subject-independent hence objective description or explanation of a part of reality that falls within the relevant scope of scientific inquiry. Classical physics, usually by definition the whole of physics accepted in the year 1900, qualifies as 'classical' in Bohr's sense. Classical physics is needed to guarantee the objectivity of QM.

All *modal interpretations* obviously qualify as interpretations of QM and are much more modest in their extensions of the vocabulary of QM_0 than Bohr cs. One modal interpretation rejects the projection postulate of QM, rejects measurement as a primitive concept in the vocabulary, makes the Evolution Postulate (**P1**) hold unconditionally, and replaces the standard property postulate with the Sufficiency Property Postulate and the

■ **BiModal Property Postulate (Dieks-Vermaas).** *The subsystems of a composite system have one of the quantitative properties $\langle B, b \rangle$, such that the basis of the Schmidt bi-orthogonal decomposition of the state of the composite system is the eigenbasis of B, with probability as in the Probability Postulate* (**P3**).

The Everett interpretation also qualifies as an interpretation of QM because it changes the vocabulary of QM_0 (adding the concept of a *branch*, or a *perspective*, or a *world*) and adds a branching postulate:[13]

■ **Branching Postulate (Everett).** *Consider a particular basis of the Hilbert-space $\mathcal{H}$, associated with physical system S, and expand its physical state $|\psi\rangle \in \mathcal{H}$ (State Postulate) in this basis, say $|\phi_j\rangle \in \mathcal{H}$, for $j = 1, 2, \ldots \dim(\mathcal{H})$. Then relative to branch j, S has physical property $\langle B, b_j \rangle$, where $B|\phi_j\rangle = b|\phi_j\rangle$.*

A solution of the problem *which* basis to consider is nowadays sought by an appeal to 'decoherence', which is the generic phenomenon that when a physical system S is in a physical environment (radiation, heat bath, air), the state becomes diagonal in some particular basis, the 'decoherence basis'. Often this basis corresponds to the physical magnitude energy or position, and it is this basis, so to speak 'preferred by Mother Nature', which then is considered in the Everett Postulate above, notably by Oxonian Everettians. They thus have physical reasons to attach ontological significance to the terms of $|\psi\rangle$ when expanded in one basis rather than an infinitude of other bases — perhaps even excellent physical reasons —, but that does not mean that they do adere ontological significance to *these* terms, and that means that QM_{Ev} goes above and beyond QM_0, — and, of course, differs from QM_{ox}.

Even Bohmian Quantum Mechanics (BQM) qualifies as an interpre-

---

[13]Our approach belies the often heard phrase that Everettian QM *is* QM, or *is* 'the quantum-mechanical formalism without any interpretation'. Dream on.

tation of QM. BQM adopts the Hilbert-space of complex wave-functions on configuration space. For the sake of simplicity, we consider 2 spinless particles in 3-dimensional space, having mass m_1 and m_2. The wave-function of the composite system is: $L^2(\mathbb{R}^3) \otimes L^2(\mathbb{R}^3) \simeq L^2(\mathbb{R}^6)$.

■ **Bohmian State Postulate.** *The state of this 2-particle system is represented by:* $\langle \psi, \mathbf{Q} \rangle$, *where* $\psi : t \mapsto \psi(t) \in L^2(\mathbb{R}^6)$ (**P0. Hilbert-Space Postulate**) *and* $\mathbf{Q} : t \mapsto \mathbf{Q}(t) \in \mathbb{R}^6$ (**Position Postulate**: see below.)

Just as in QM_0, ψ is postulated to obey the Schrödinger equation. Vector $\mathbf{Q}(t)$ consists of 6 components, and can be written as $\langle \mathbf{Q}_1(t), \mathbf{Q}_2(t) \rangle$, where $\mathbf{Q}_1(t), \mathbf{Q}_2(t) \in \mathbb{R}^3$. Vector $\mathbf{Q}_1(t)$ represents the position of $\mathbf{p}_1$ at time t, and similarly $\mathbf{Q}_2(t)$. Like in classical mechanics but unlike in QM_{ox}, in BQM every particle always has a position. Bohmians 'complete' QM by adding $\mathbf{Q}$ to ψ.

■ **Position Postulate.** *The positions of the particles are determined by* ψ *via the Guiding Equation, which is for particle 1:*

$$m_1 \frac{d\mathbf{Q}_1(t)}{dt} = \hbar \, \mathrm{Im} \left(\frac{\nabla_1 \psi(\mathbf{q}_1, \mathbf{q}_2, t)}{\psi(\mathbf{q}_1, \mathbf{q}_2, t)} \right)_{\mathbf{q}_1 = \mathbf{Q}_1(t)}, \quad (5)$$

where ∇_1 is the gradient, with respect to $\mathbf{q}_1$, and $\mathrm{Im}(z) \in \mathbb{R}$ is the imaginary part of $z \in \mathbb{C}$. Similarly for particle 2.

One should not confuse $t \mapsto \mathbf{Q}_1(t)$ with $\mathbf{q}_1$: the afore-mentioned describes the path of particle 1 in 3-dimensional space, whilst the last-mentioned is a physically uninterpreted variable of ψ.

The left-hand-side of the Guiding Equation (5) is a time-derivative of the position of particle 1, which is the definition of its velocity:

$$\mathbf{v}_1^{\psi}(t) = \frac{d\mathbf{Q}_1(t)}{dt}, \quad (6)$$

where the superscript 'ψ' is there to emphasise that the velocity is determined by ψ, via eq. (5), which pertains to the composite system. There is further an ■ **Equilibrium Postulate**, which posits the Born-measure for position probabilities. From this and an elaborate story that reduces all measurements to position measurements, the Probability Postulate follows.

Legenda table below. All theories entail the postulates of QM_0, which are therefore omitted; only the additional postulates are mentioned. By '1/2' is meant the Sufficiency Property Postulate. ■ **Categorical Evolution Postulate**: always unitary evolution over time, whether measurements are performed or not.

	QM$_{ox}$	BiModQM	QM$_{Cop}$	EvQM	BQM
St. Prop. Post.	+	1/2	1/2	1/2	−
Pure State Post.	+	+	+	+	+
Projection Post.	+	−	+/−	−	−
Spectrum Post.	+	+	+	+	+
Categ. Evol. Post.	+	+	+/−	+	+
Quantum Post.	−	−	+	−	−
Buffer Post.	−	−	+	−	−
Compl. Post.	−	−	+	−	−
BiMod. Prop. Post.	−	+	−	−	−
Branching Post.	−	−	−	+	−
Bohm. State Post.	−	−	−	−	+
Position Post.	−	−	−	−	+
Equilibr. Post.	−	−	−	−	+

6 True Inwardness of Reality

What we have not done is to expound yet another interpretation of QM, to defend one or to criticise one. What we have done is something more modest. We have expounded what it means *to interpret* QM and it means, in a nutshell, this: to extend QM$_0$ by adding postulates and enriching the vocabulary. This is achieved by proceeding as follows.

❶ List the concepts that the interpretation employs *in addition to* those of minimal QM (QM$_0$), which are: physical system, physical subsystem, physical magnitude, probability, measurement; explain these additional concepts.

❷ Mention whether the physical concepts of QM$_0$ change in the new interpretation, i.e. whether the meaning of the words expressing them differs in the new interpretation when these words are already employed in QM$_0$; explain these differences.

❸ Mention whether the logical concepts of QM$_0$ change in the new interpretation, i.e. whether they deviate from those expressed in classical logic, and explain these differences.

❹ List the postulates that the new interpretation adds to those of QM$_0$; if postulates of QM$_0$ are not among those of the new interpretation, show that these postulates of QM$_0$ become theorems in the new interpretation.

❺ Mention whether some (or all) of the postulates of QM_0 change in the new interpretation; explain these changes.

❻ List the questions that QM_0 does not answer, or the problems that QM_0 does not solve, and show how the new interpretation of QM answers (some of) them and solves (some of) them, respectively.

The above list ought to be the to-do list for every interpreter of QM.

Thus the interpretation of QM turns out to be *not the same as* how 'to interpret' is generally interpreted in philosophy, which is: *to assign meaning to*. Depending on the interpretation under consideration, there is more or less of interpretation in the last-mentioned sense going on; but the thesis that this is *all* that is going on in the discourse on the interpretation of QM is like saying that arranging the table is all that is going on in the preparation of a dinner.

Is the interpreation of QM, then, perhaps a special case of hermeneutics as we have come to know it in continental philosophy, where we think of the likes of Schleiermacher, Dilthey, Heidegger, Gadamer and Derrida? Is the discourse on the interpretation of QM an hermeneutical discourse in his sense, i.e. is there such a thing as *quantum hermeneutics*? Let us briefly take a closer look at hermeneutics in philosophy.

The word 'hermeneutics' comes from the Greek word for interpretation or translation ($ερμηνευω$), which derives from the name of the Greek mythological figure Hermes, who deciphered messages of the gods and communicated them to human mortals. Aristotle introduced hermeneutics in philosophy in his *De Interpretatione*, by distinghuishing the symbols or signs (*symbola*) from the affect they have on our minds (*pathemata*) as well as from the entities they represent (*pragmata*), of which the mental affections are representations (*homoiomata*). Hermeneutics in philosophy is the study of written texts in context, in order to understand the text better, to get more grip on its content. The study of sacred texts in Talmudic, Vedic, Biblical and Apostolic traditions belong to *theological* or *religious* hermeneutics; they have one leg in mythology (Hermes) and the other one in philosophy (Aristotle). The context of the text is usually taken to be the historical context in which the text is produced (Dilthey), in order to understand the views and intentions of the author (Schleiermacher) or to understand what the text itself expresses (Dilthey), where 'understanding' has to be understood in the sense of Droysen's *verstehen* rather than *erklären*.[14] Heidegger gave birth to *existential hermeneutics*, an endeavour to understand human existence, *Dasein*, directly, without mediation by text and language

---

[14]Wright [1971], p. 5.

generally. Inquiry into written text in context, call it *textual hermeneutics*, is something else: indirect and further removed from life as we live and experience it. Existential hermeneutics was further developed by Heidegger's pupil Gadamer [1960], who further delved into individual human experience, mediated by language however, in particular by spoken language in conversation ('to experience truth').

The *hermeneutic circle*, an idea introduced by Heidegger, has various manifestations. One is that in order to understand parts of a text, one needs to understand the text as a whole, and in order to understand the whole text, one needs to understand its parts. The process of interpretation, leading to an ever increasing understanding, thus proceeds in a circle of reading and re-reading. One understands the postulates of QM better after one has understood the whole of QM, and one understands the whole of QM better after one has understood the postulates. This is however not what is going on in the discourse of the interpretation of QM. Another manifestation of the hermeneutic circle is the reciprocity between text and context. But inquiry into the historical context of the advent of QM, and into what the effect of the historical context on the content of quantum-mechanical texts has been belong to the discourse of the history of QM, not to the interpretation of QM. So this second manifestation of the hermeneutical circle also is definitely not what is going on in the discourse of the interpretation of QM.

Derrida took a turn in textual hermeneutics by considering only *other texts* as the context of a text, leading to his notorious assertion "There is nothing outside the text."[15] "There is nothing outside context", expresses the same, Derrida later explained.[16] Notice that we only have access to the past, to the factual historical context in which a text is written, by means of other texts — and occasionally images and artefacts. Use of words in other texts resonate in the text under consideration, and their use in the text under consideration resonate back in all other texts. This seems yet another manifestation of the hermeneutic circle. But, again, this hardly helps to capture what is going on in the interpretation of QM.

Parenthetically, in *Finnegans Wake* Hermeneutic Circles are Everywhere (HCE).

We tentatively conclude that there is no such thing as 'quantum hermeneutics'.[17] This conclusion savours an *a priori* possible and per-

---

[15]Derrida [1967], pp. 158–159. Derrida was parenthetically heavy influenced by *Finnegans Wake*, see Derrida [1984].

[16]Derrida [1988], p. 148.

[17]In the sense of hermeneutics as understood in the philosophical tradition sketched above. In another, literal sense, 'quantum hermeneutics' just means 'quantum inter-

haps promising connexion between the discourse of (i) philosophy of physics and of (ii) hermeneutics in philosophy — and philosophy of language we submit. Interpreting QM is not merely 'a matter of semantics' or penetrating deeper into quantum-mechanical texts and their context. What is at stake in the discourse of the interpretation of QM is *how* and *what* microphysical reality is, *how to understand* microphysical reality — if it is understandable by us at all —, granted that QM provides us with the best basis to answer these questions. What is at stake here are answers to all sorts of questions concerning microphysical reality, the world of the tiny and the brief, and to physical reality generally. Hopefully the answers to these questions jointly provide some coherent understanding of physical reality. Finding answers becomes a matter of finding the right additional postulates to extend QM_0, rather than just keeping the postulates fixed and re-interpreting expressions occurring in them. Novel concepts, alien to Nalasweknowit, not in use and nowhere expressed in other texts, may very well have to be constructed for this purpose. Steps ❷–❺ are supposed to involve precisely radical conceptual change.

Hence in one of the most successful areas of natural science, quantum physics, an interpretational inquiry was launched by theoretical physicists in the 1920ies; later philosophers joined in, with a vengeance. A mainstream interpretation was settled, of Copenhagen design. But it did not last. Copenhagen QM has left too many questions unanswered. Schrödinger complained that the interpretational problems of QM were shelved, not solved. In his Nobel Lecture of 1969, Murray Gell-Mann notoriously declared that an entire generation of physicists was brainwashed into believing that the interpretation problems of QM were solved, by the Great Dane. To interpret QM is to extend QM_0 and its vocabulary, which permits the expression of more concepts than the language of QM_0 permits. Since forging novel concepts is a philosophical activity *par excellance*, a philosopical activity is required to aid physics to achieve its aims.

We want to unveilop the theory of the building blocks of matter and their interactions. We need to destructify the obscuritads it bangcreated, in order to sunshine physical reality by a syntasm of ison and remagination. For we want to understand. We need to understand. We shall understand. In our stream of consciousness, from swerve of shore to bend of bay, we want to float and fly, transpicuously and persparantly, and not to drown and die.

---

pretation'.

The final word is to B.C. van Fraassen, with an empiricist twist at the end:

> Why then be interested in interpretation at all? If we are not interested in the metaphysical question of what the world is really like, what need is there to look into these issues?

> Well, we should still be interested in the question of how the world could be the way quantum mechanics — in its metaphysical vagueness but empirical audacity — says it is. That is the real question of *understanding*. To *understand* a scientific theory, we need to see how the world could be the way that the theory says it is. An *interpretation* tells us that. The answer is not unique, because the question 'How could the world be the way the theory says it is?' is not the sort of question to call for a unique answer. Faith in the actual truth of a good answer, so interpreted, is neither required by *understanding*, nor does it help.[18]

7 Patrick Colonel Suppes

Permit me to admit that it feels awkward to contribute to an edited volume for Patrick Colonel Suppes (1922–2014). For Suppes has been suspiciously silent in all of his writings about the interpretation of QM. Did he fancy the Copenhagen Interpretation? Did he prefer some modal interpretation? Was he an Everettian in cloack and dagger? No favourites at all? Lame agnosticism? Brute rejection of the very issue of the interpretation of QM as a pseudo-issue? Was he aware of the obscuritads that circumveilop us?

Suppes has not been entirely silent about QM. Everything he has written about QM is about or connected to *probability* (available at his website of Stanford University, spanning *seven* decades).[19] Concerning the issue of the *the interpretation of probability*, however, in and outside QM, Suppes has also been suspiciously silent about where his sympathy lies, his insistence that the differences in interpretation of probability ought to be characterised mathematically notwithstanding.[20] Was he a devout Bayesian? Or some frequentist? Did he believe in propensities?

If there were answers, they have been blown away by the winds of immortality.

Acknowledgments.

---

[19]Suppes [1963], [1965].
[20]Suppes [2001], Chapter 5.

Thanks to Gijs Leegwater, Geurt Sengers and Stefan Wintein (Erasmus University Rotterdam), and an anonymous Referee for comments.

BIBLIOGRAPHY

[1] P. Bush, P.K. Lahti, and P. Mittelstaedt, *The Quantum Theory of Measurement* (2nd. Rev. Ed.). Berlin: Springer-Verlag, 1996.
[2] J. Derrida, *Of Grammatology*, translated from the French original of 1967 by G.C. Spivak. Baltimore & London: Johns Hopkins University Press, 1976.
[3] J. Derrida, 'Two words for Joyce'. In: *Post-structuralist Joyce: Essays from the French*, D. Attridge and D. Ferrer (eds.). Cambridge: Cambridge University Press, pp. 145–160, 1984.
[4] J. Derrida, *Limited Inc.*. Evanston: Northwestern University Press, 1988.
[5] P.A.M. Dirac, *The Principles of Quantum Mechanics*. Cambridge: Cambridge University Press, 1928.
[6] M.A.E. Dummett,*The Logical Basis of Metaphysics*. Cambridge, Massachusetts: Harvard University Press, 1991.
[7] B.C. van Fraassen, *Quantum Mechanics: An Empiricist View*. Oxford: Clarendon Press, 1991.
[8] H.-G. Gadamer, *Wahrheit und Methode: Grundzüge einer philosophischen Hermeneutik*. Tübingen: Mohr, 1960.
[9] J. Joyce, *Finnegans Wake*. London: Faber and Faber, 1939.
[10] F.A. Muller, 'The Deep Black Sea: Observability and Modalify Afloat', *British Journal for the Philosophy of Science* **56**, pp. 61–99, 2005.
[11] J. von Neumann, *Mathematische Grundlagen der Quanten Mechanik*, Berlin: Springer-Verlag., 1932.
[12] S.W. Saunders, 'Time, Quantum Mechanics, and Tense', *Synthese* **107**, pp. 19–53, 1996a.
[13] S.W. Saunders, 'Naturalizing Metaphysics', *The Monist* **80**, pp. 44–69, 1996b.
[14] P.C. Suppes, 'The role of probability in quantum mechanics', in: *Philosophy of Science: the Delaware Seminar*, Volume 2, B. Baumrin (ed.), New York: Wiley & Sons, pp. 319–337, 1963.
[15] P.C. Suppes, 'Probability Concepts in Quantum Mechanics', *Philosophy of Science* **28.4**, pp. 378–389, 1965.
[16] P.C. Suppes, *Representation and Invariance of Scientific Structures*. Chicago: University of Chicago Press, 2001.
[17] I. Votsis, 'Perception and Observation Unladened', *Philosophical Studies* **172**, pp. 563–585, 2015.
[18] D. Wallace, *The Emergent Multiverse. Quantum Theory according to the Everett Interpretation*, Oxford: Oxford University Press, 2012.
[19] G.H. von Wright, *Understanding and Explanation*, London: Routledge & Kegan Paul, 1971.

Structures in Science and Metaphysics
NEWTON C. A. DA COSTA AND OTÁVIO BUENO

ABSTRACT. In this paper, we develop a formalization of a metaphysics of structures and provide new applications to geometry and to metaphysics. We adopt the theory of set-theoretical structures formulated in da Costa and French [15] and in da Costa and Rodrigues [16]. The applications are made in the context of extending the proposal advanced in Caulton and Butterfield [11] to a higher-order metaphysics.

Introduction

This paper has three main purposes: (1) It aims to discuss some aspects of the general theory of set-theoretic structures as presented in da Costa and French [15]. However, our exposition is in principle independent of the contents of this book, being in good part based on the paper da Costa and Rodrigues [16]. The theory of structures studied here can be employed in the axiomatization of scientific theories, particularly in physics. The theory is related to that of Bourbaki [4], but our emphasis is on the semantic dimension of the subject, which is not in agreement with Bourbaki's syntactic views. Set-theoretic structures are important, among other things, because they constitute the basis for a philosophy of science, as it was delineated in da Costa and French [15] (see also Suppes [34]). After developing the framework, we examine some aspects of geometry in order to make clear how the central notions of the theory can be applied to higher-order structures found in science. The case of geometry is significant, since pure geometric structures can be considered not only as abstract mathematical constructs, but also as essential tools for the formulation of physical theories, in particular the theories of space and time.

(2) The paper also aims to discuss how to extend ideas of Caulton and Butterfield [11] to what may be called higher-order metaphysics. In this metaphysics, in addition to first-order objects, we also find their corresponding properties and relations of any type in a convenient type hierarchy.

(3) Finally, the paper outlines the first steps of the formalization of a metaphysics of structures, or structural metaphysics, inspired by ideas of authors such as French and Ladyman (see, for example, Ladyman [29], French and Ladyman [19], and French [18]). This kind of metaphysics may be seen as an adaptation of the higher-order metaphysics of structures implicit in Caulton and Butterfield [11].

1 Set theory, structures, and languages

In this section we present the set-theoretic framework that will be applied subsequently in the paper, which summarizes some concepts and results of da Costa and Rodrigues [16] and da Costa and Bueno [13]. All our (set-theoretic) constructions are implemented in Zermelo-Fraenkel set theory, which is the basis for the framework (in particular, we take languages to be a certain kind of free algebra).

The set T of types is defined as follows:

1. The symbol i belongs to T;
2. If $t_0, t_1, \ldots, t_{n-1} \in T$, then $\langle t_0, t_1, \ldots, t_{n-1}\rangle$, $1 \leq n < \omega$ also belongs to T.
3. The elements of T are only those given by clauses 1 and 2.

The order of a type t, $ord(t)$ is introduced as follows:

1. $ord(i) = 0$.
2. $ord(\langle t_0, t_1, \ldots, t_{n-1}\rangle) = max\{ord(t_0), ord(t_1), \ldots, ord(t_{n-1})\} + 1$

The transitive closure $[t]$ of a type t is given by the clauses:

1. $t \in [t]$.
2. If $t = \langle t_0, t_1, \ldots, t_{n-1}\rangle$, then $t_0, t_1, \ldots, t_{n-1}$ belong to $[t]$.
3. The elements of $[t]$ are only those given by clauses 1 and 2.

We have: $[t] = [t_0] \cup [t_1] \cup \ldots \cup [t_{n-1}]$. All elements of $[t]$, different from t, have orders strictly less than the order of t. On the other hand, $[i] = \emptyset$ and $ord(i) = 0$. There is a partial ordering in the set of types, $\leq$, defined by the condition: $t_1 \leq t_2$ if $t_1 \in [t_2]$. Any decreasing sequence of types is finite, i.e., the set of types is *regular*.

If D is a set, then we define a function τ_D or, to simplify, τ, whose domain is T, by the following clauses:

1. $\tau(i) = D$.
2. If $t_0, t_1, \ldots, t_{n-1} \in T$, then $\tau(\langle t_0, t_1, \ldots, t_{n-1}\rangle) = \mathcal{P}(\tau(t_0) \times \tau(t_1) \times \ldots \times \tau(t_{n-1}))$, where $\mathcal{P}$ and $\times$ are the symbols for the power-set and the Cartesian product, respectively.

The set $\bigcup range(\tau_D)$ is denoted by $\varepsilon(D)$, and is called the scale based on D. The objects of $\tau(i)$ are called individuals of $\varepsilon(D)$, and the objects of $\tau(t)$, $ord(t) > 0$, are called objects or relations of type t; the type of individuals is i. We also have:
$$\hat{\varepsilon}(D) = \bigcup(range(\tau_D) - \{D\}).$$

The cardinal $k(D)$, defined by the condition

$$k(D) = sup\{\overline{\overline{D}}, \overline{\overline{\mathcal{P}(D)}}, \overline{\overline{\mathcal{P}(\mathcal{P}(D))}} \ldots\},$$

is the cardinal associated with $\varepsilon(D)$. (Sometimes, instead of $k(D)$, we write k_D.)

A sequence is a function whose domain is an ordinal number, finite or infinite. $\hat{b}_\lambda$ is the range of the sequence $\hat{b}$, and λ is an ordinal.

We call structure e with basic set D an ordered pair

$$e = \langle D, r_\iota \rangle$$

where r_ι is a sequence of elements of $\varepsilon(D)$. We also call D the domain of e, and it is supposed to be non-empty. D and the terms of r_ι are the primitives elements of e; the elements of D are the individuals of e. We shall identify such individuals with their unit sets when there is no danger of confusion. $\varepsilon(D)$ and $k(D)$ are the scale and the cardinal associated to e, respectively. In all cases, the ordinal which is the domain of r_ι is strictly less than $k(D)$. $\hat{\varepsilon}(D)$ is the strict scale associated to e, and contains all relations of $\varepsilon(D)$, including the unary relations, that is, sets.

The order of an object of $\varepsilon(D)$ is the order of its type. The order of $e = \langle D, r_\iota \rangle$, denoted by ord($e$), is defined as follows: if there is the greatest order of the objects of the range of r_ι, then ord(e) is this greatest order; otherwise, ord(e) = ω.

$L^\omega_{\omega k}(R)$ is the higher-order, infinitary language introduced in da Costa and Rodrigues [16]. Its blocks of quantifiers are finite, and conjunctions and disjunctions have length strictly less than $k(D)$, when $k(D) > \omega$; if $k(D) = \omega$, such conjunctions and disjunctions are always finite. R is the set of constants of the language, each having a fixed type.

$L^\omega_{\omega k}(R)$ can be interpreted in structures of form

$$e = \langle D, r_\iota \rangle,$$

the constants denoting D and the objects r_ι, each constant possessing the same type as that of the object it denotes. The sequence r_ι has as its domain an ordinal strictly less than $k(D)$, usually finite or denumerable. In what follows we may identify a constant with the object it denotes; this will be done when there is no danger of confusion. In the structure $e = \langle D, r_\iota \rangle$, D and r_ι are called the primitive concepts of e.

Sentences of our languages are formulas without free variables. We say that a sentence ξ of a language, interpreted in a structure

$$e = \langle D, r_\iota \rangle$$

is true in this structure in analogy with the case of first-order (finitary) logic. To express that ξ is true in e we write:

$$e \models \xi.$$

Notions such as model, definability, etc. are defined as in da Costa and Rodrigues [16].

From now on, we usually suppose that in the structure $e = \langle D, r_\iota \rangle$ the family r_ι of relations is finite (i.e., the domain of r_ι is a finite ordinal). When r_ι is infinite this will be explicitly stated. As a consequence, every structure will have, usually, an order that is a positive integer or zero.

To each structure of order n, $e = \langle D, r_\iota \rangle$, it is associated a finitary language of order n $L^n_{\omega\omega}(R)$ and an infinitary language $L^n_{\omega k_D}(R)$, or to simplify $L^n_{\omega k}(R)$, where R is the range of r_ι and whose variables have types of order at most n. These languages are the strict languages associated with the structure e of order n.

The kind of structure $e = \langle D, r_\iota \rangle$ is the sequence t_ι, where t_ι is the type of r_ι. The two languages associated with a given structure are both interpretable in this structure. But, for convenience, isomorphic structures of the same kind will be identified, despite the obvious differences between them.

In the languages associated with a structure $e = \langle D, r_\iota \rangle$ of order n, we may add variables of any order m, with $m > n$; in particular, m may be ω. Clearly such languages are interpretable in $e = \langle D, r_\iota \rangle$, via $\varepsilon(D)$. In this case, e becomes the basis of a semantics of those languages. In particular, Peano arithmetic is a first-order structure, but there are corresponding arithmetic theories of any order whatsoever.

If $e = \langle D, r_\iota \rangle$ is a structure of order n, $e^{(m)} = \langle D, s_\lambda \rangle$ will be the structure in which s_λ is the sequence of all relations of $\varepsilon(D)$ of order $\leq m$ definable in the wide sense in $e = \langle D, r_\iota \rangle$ by the means of the infinitary language associated with e. Sometimes, all the relations s_λ are definable in the wide sense in terms of a finite number of relations in the range of s_λ. Thus, it is possible to take s_λ as a finite sequence. $e^{(m)}$ constitutes the m-order counterpart of e, and the m-order semantics of e is the semantics of $e^{(m)}$. For example, although elementary Euclidean geometry is not a first-order structure, it is possible to investigate its first-order counterpart (see Tarski [35]).

Despite the fact that the language of set theory is first-order, structures of any order whatsoever can be studied in it and the corresponding theory of sets. This is, of course, a different situation from the case of elementary Euclidean geometry just mentioned, which involves the study of the first-order counterpart of a non-first-order structure. The semantic interplay between a language (that is, a free algebra) and a structure can be treated by means of set theory (say, Zermelo-Fraenkel set theory) even when the structure is of order $n > 1$.

The theory of the kind of structure of $e = \langle D, r_\iota \rangle$, or the species of structures to which e belongs, can be formulated in one of its associated languages, sometimes enlarged by extra variables and constants of higher-order types, or in the language of set theory. This is accomplished by the choice of appropriate postulates, formulated in the chosen language.

Theories, species of structures or axiomatic systems are formulated as follows: (1) It is given a language, which may be that of set theory extended by the addition of new primitive terms, or associated with structures of a certain kind. (2) The postulates and rules of the underlying logic are those naturally connected with the language of clause (1) (in the case of set theory, employed as the underlying framework, its axioms are clearly included among the postulates of the theory). (3) Specific postulates for the theory are also introduced. (4) By means of the logic of clause (2) and the specific axioms, the concept of theorem is defined.

In this way, one immediately obtains the common syntactic and semantic results regarding theories. Our notion of theory, axiomatic system or species of structures is, therefore, more general than the one articulated in Bourbaki [4], since we are not restricted to syntactic considerations alone.

Two important concepts are those of transportable formula and of intrinsic term that we will introduce informally. The postulates of a theory are supposed to be transportable formulas. A formula of the language of theory T, interpretable in structures of the kind of the structure e, is transportable if, and only if, being true in one structure of this kind it is also true in every structure isomorphic to it.

Terms may be joined to the language of a theory as new constants or with the help of the description operator. We say that term t is intrinsic in a theory T if, given any two isomorphic models e and e' of T, then for any isomorphism ξ between e and e', the denotation of t in e' is $\xi(a)$, where a is the denotation of t in e.[1]

These considerations can be extended to the case in which the starting point is a finite sequence $D_1, D_2, \ldots D_m$ of basic sets and the relations r_ι involve elements of the scale $\varepsilon(D_1 \cup D_2 \cup \ldots \cup D_m)$. The notions of type, order of a type, scale, individuals, etc. are easily adapted to this new case. A structure, then, is a set-theoretic construct

$$e = \langle D_1, D_2, \ldots D_m, r_\iota \rangle,$$

where $D_1, D_2, \ldots D_m$ is the finite sequence of basic sets and r_ι are the primitive relations of e. The sequence $D_1, D_2, \ldots D_m$ (or, to abbreviate, D_δ) is composed by arbitrary (non-empty) sets, and the notions of language associated with e, theory or species of structures of the kind of e, etc. are introduced without difficulty. We usually say that the sets D_δ are endowed with the species of structures of e.

In most cases, some or all basic sets are supposed to be endowed with previously defined structures (for example, in the species of real vector spaces, the scalars are the real numbers); these basic sets are called auxiliary sets. The other basic sets constitute the strict basic sets. Normally, structures have at least one strict basic set.

If the family r_ι in the structure $e = \langle D, r_\iota \rangle$ is empty, then e reduces to the bare set D.

An isomorphism between two structures of the same kind $e_1 = \langle D_\delta, r_\iota \rangle$ and $e_2 = \langle B_\delta, s_\iota \rangle$ is a family of bijections between D_δ and B_δ satisfying obvious conditions, plus the following: the bijection between two corresponding auxiliary sets is, for simplicity, the relation of identity. In other words, isomorphisms must keep invariant the corresponding auxiliary sets and their associated structures.

---

[1] The notions of transportable formula and of intrinsic term were introduced by Bourbaki [4] (compare with da Costa and Rodrigues [16], and da Costa and Chuaqui [14]).

Therefore, the structure

$$e = \langle D_1, D_2, \ldots D_m, r_\iota \rangle$$

may be written as follows:

$$e = \langle D_\delta, A_\delta, r_\iota \rangle$$

where D_δ is the family of strictly basic sets and A_δ is the family of auxiliary sets.

Structures are commonly presented as finite sequences of previously defined structures arranged by appropriate relations among them. So, the notion of species of structures (theory or axiomatic system) has to be adjusted to the new situation. But that is not difficult to be done from the conceptual point of view.

For instance, a Riemannian geometric structure (Riemannian space or Riemannian geometry) is an ordered pair

$$R = \langle D, m \rangle,$$

where D is a (real) differential manifold and m is a Riemannian metric on D (that is, a symmetric, positively defined, 2-covariant tensor field on D). The theory of Riemannian spaces has as its models structures such as $R = \langle D, m \rangle$.

Another sort of structures, usually occurring in physics, is that of a bundle. A bundle is a triplet

$$F = \langle E, \pi, D \rangle$$

where E (the bundle space) and D (the base space) are topological spaces, and π (the bundle projection) a continuous map of E on D.

Structures presented as sequences of other structures are reducible to structures in the standard sense, although it must be clear what are the basic sets and the auxiliary sets. The reduction is expressed, among other things, by the description of the inter-relations imposed on the component structures. For example, in the case of bundles, the reduction to the standard version may be as follows:

$$F = \langle A, B, t_A, t_B, s \rangle$$

where t_A and t_B are collections of subsets of A and B, respectively, and s is a function of A on B. We then introduce the postulates of F, which are the following ones:

1. t_A is a topology on A;

2. t_B is a topology on B;

3. s is a continuous function of the topological space $\langle A, t_A \rangle$ on the topological space $\langle B, t_B \rangle$.

The reduction of a Riemannian structure to its standard formulation would be more complex, but it can still be formulated.

2 Definability and expressibility

In some cases, to simplify the exposition, we identify relations with their names. In the structure $e = \langle D, r_\iota \rangle$, we say that $s \in \varepsilon(D)$ is strictly definable in e if the following condition is met: there is a formula $F(x)$ of the finitary language associated with e, in which x is its sole free variable, such that:

$$e' \models \forall x(x = s \leftrightarrow F(x))$$

in the language just mentioned and to which the new primitive term s was added, where e' is the structure e with the relation s included.

Let $e = \langle D, r_\iota \rangle$ be a structure, $\varepsilon(D)$ its scale, s an element of $\varepsilon(D)$, and l_μ ($\mu < k_D$) a sequence of objects of $e = \langle D, r_\iota \rangle$. We then define, by induction, that the expression s is expressible in the sequence l_μ in structure e' as follows:

1. if s is definable in the strict sense in the structure e to which we have joined the terms of l_μ as new primitive relations, then s is expressible in the sequence l_μ in e;

2. if s is not an individual of e, then s is expressible in l_μ in e if every element of s is expressible in l_μ in e;

3. if s is expressible in the sequence b_α of elements of $\varepsilon(D)$ in the structure e and every element of b_α is expressible in l_μ in e, then s is expressible in l_μ in the structure e.

Expressibility is equivalent to definability in the infinitary language associated with e extended by the addition of the term s (see da Costa and Rodrigues [16]). Instead of 'expressibility', we also use the expression 'definability in the wide sense'.

A definition of the new constant c joined to the language of the theory T is an expression of the form $c = \iota x F(x)$, such that:

$$T \vdash \exists ! x F(x),$$

where $F(x)$ is a formula of the original language of T (without c) with x as its only free variable.

The previous definition can be adjusted to the cases in which the constant c belongs to the language of T, or in which T has no specific postulates (such postulates constitute the empty set), or when the notion of semantic consequence ($\models$) is used instead of that of syntactic consequence.

Let L be a language, $e_1 = \langle D, r_\iota \rangle$ be a structure whose language $L(D, r_\iota)$ is an extension of L by the introduction of the new constant D and the family r_ι of new primitive constants, and $e_2 = \langle D, s_\lambda \rangle$ be another structure having its language defined analogously. In this case, e_1 is said to be strictly (widely) equivalent to e_2 if every s_λ is definable in the strict (wide) sense in terms of D and r_ι, and conversely. For example, in the language of set theory, two topological structures, $t_1 = \langle D, v \rangle$ and $t_2 = \langle D, A \rangle$, characterized by the family of neighborhoods and the set of open sets, respectively, are strictly equivalent. In the same language, the structure of natural numbers, in the sense of von Neumann, is widely equivalent to the structure of real numbers, conceived as Cauchy sequences of rational numbers; however, there are real numbers that are not strictly definable in terms of natural numbers, although they are widely definable in terms of these numbers.

L is the language of set theory. We assume that the postulates of L are the common postulates of ZF (Zermelo-Fraenkel set theory). Suppose that T_1 is a theory whose language is $L(D, r_i)$, where D denotes a set and r_i is a family of relations over D. We denote by T_2 another theory whose language is (LE, s_λ), similarly defined. Under these conditions, T_1 and T_2 are equivalent if:

1. every s_λ and E are definable by formulas of $L(D, r_\lambda)$ in T_1, and every r_i and D are definable in $L(E, s_\lambda)$ by formulas of $L(E, s_\lambda)$ in T_2;

2. every postulate of T_2, reformulated in the language of T_1, is provable in T_1, and conversely.[2]

The usual metamathematical results involving mathematical structures and species of structures can be obtained in this approach. For instance, we have: (1) Any two models of second-order arithmetic are isomorphic. (2) First-order arithmetic is not categorical. (3) In first-order languages, as usually formulated (with a first-order semantics), the notion of finite group is not axiomatizable. (4) In a convenient infinitary, first-order arithmetic, all subsets of natural numbers are widely definable. (5) In set theory, any structure of any order can be extended to a rigid structure, by adding a finite number of new relations (da Costa and Rodrigues [16]).

3 Theories and operations with structures

As noted, structures are correlated with, at least, three classes of language: the two classes associated with them and that of set theory. The orders of the languages associated with a structure e, languages that can be finitary or infinitary, are the same as that of e. Nonetheless, as also noted, any structure can be studied in set theory, i.e., with the help of the tools of set theory, whose language is first order. We need to be careful here, since a given structure of order n has an intended semantics of order n. In a certain sense, the order of structure e is kept invariant in set theory, but the membership relation strongly enriches the intended semantics of e. Concepts like definability, definability in the wide sense, discernibility, indiscernibility, individuality, etc., to be examined below, are essentially language dependent.

From now on, we reserve the expression 'species of structures' for theories axiomatized in set theory (ZF with its axioms), but whose language is extended by adding extra symbols denoting relevant sets and relations. When a different language is employed, this will be made explicit.

Species of structures have important features. We highlight some below:

1. When a species of structures is based on various sets, it is easy to transform the species into an equivalent one with only one basic set. However, if the underlying language of a theory differs from that of set

---

[2]There are some obvious variations in the definition of theory equivalence. But we need not discuss them here (see, for instance, Shoenfield [33] for the case of finitary, first-order languages).

theory, for example, if it is a first-order language involving various specific primitive symbols (constants), this transformation is not always possible. To implement the transformation, we obviously need the resources of set theory. Shoenfield [33], for instance, axiomatizes what is essentially second-order arithmetic in a two-sorted, first-order language. But it is not possible to reduce the basic sets to only one in the language used. Some non-standard models are also included via the corresponding first-order semantics.

2. The usual frameworks for the study of higher-order structures are set theory or higher-order logic. In both cases, the intended, standard semantics is adopted. However, a many-sorted, first-order logic can also be employed, with the inclusion of first-order structures.

3. The specific axioms of a species of structures are always supposed to be transportable and its terms intrinsic. When the underlying logic of a theory is type theory (higher-order logic) these conditions are automatically satisfied (see da Costa and Chuaqui [14]).

4. Starting with initial structures, new structures can be obtained in set theory via various methods. The following are worth mentioning: axiomatization, (structural) deduction, combination, structural derivation, and the use of universal mappings (see Bourbaki [4]). We consider each of them in turn.

Axiomatization is the general way to introduce new species of structures. This is done, as indicated above, in accordance with the definition of species of structures. When other languages are used, the process is similar, but with suitable adjustments.

As an illustration, here are some examples of species of structures: (i) a *groupoid* is a set endowed with a binary operation; (ii) a *semigroup* is a groupoid whose operation is associative; (iii) a *semigroup* with identity is a manoid. Moreover, species of structures such as topological space, lattice, Boolean algebra, loop and topological group can also be defined.

Deduction of species of structures (or *structural deduction*) consists in a method of formulation of new structures that is best illustrated by an example (for a detailed description, see Bourbaki [4]). The real projective plane depends on the species of real numbers:

$$\mathcal{R} = \langle R, +, \times, 0, 1, \leq \rangle,$$

which is a higher-order structure. As is well known, it can be characterized as follows: (1) The starting point is a set of triples of real numbers. (2) Projective points are then introduced as classes of equivalence of triples. (3) Finally, primitive concepts of the real projective plane are introduced: the set of points (P), the set of lines (L), and the relation of incidence I (if a point and a line are incident, the point lies on the line or the line passes through the point). The structure of a real projection plane is, thus, a triplet:

$$P = \langle P, L, I \rangle$$

satisfying proper axioms. In other words, the real projective plane was obtained, by deduction, from $\mathcal{R}$ through the intrinsic terms P, L and I (whose intrinsic nature can be easily verified). It is clear that the notation:

$$\langle P, L, I \rangle$$

for the real projective plane is no more than an abbreviation, omitting many details; nonetheless, it is a useful notation.

Numerous species of structures are obtained via *combination* of other species. For example, a topological group G involves the species of structures of group and that of topological space:

$$G = \langle A, B \rangle,$$

where A is a group and B a topological space, the operations of A being continuous relatively to the topology of B. Various species of structures are obtained in this way, such as differentiable manifold, analytic manifold, and Grassmann algebra.

The operation of *structural derivation* (of structures) is related to the notion of morphism, as discussed in Bourbaki [4]. Examples of structures of this sort are the inverse image structure, induced structure, product structure, direct image structure, and quotient structure.

Free structures are introduced with the help of *universal mappings*. For example, mathematically speaking, a formal language is a kind of free algebraic structure.[3]

4 The Erlangen Program and some of its extensions

Klein's Erlangen Program is well known in geometry. Veblen, for instance, summarizes it as follows:

> The system of definitions and theorems which expresses properties invariant under a given group of transformations may be called, in agreement with the point of view expounded in Klein's Erlangen Program, *a geometry*. (Veblen and Young [36], volume I, p. 71.)

Reflecting on the concept of *space*, Wilder notes:

> In modern mathematics, the term 'space' has extremely broad connotations, the difference between 'set' and 'space' often being very slight, a 'space' being simply a set to which certain special properties have been added. The most common property is that having a *metric*, or distance function. (Wilder [38], p. 177.)

He then continues:

---

[3] A natural way of formulating the concepts of species of structures and of structure is to use category theory (see Corry [10]). But it would take us too far afield to examine this issue here.

And, according to the point of view proposed by Klein in 1872, one may speak of the ξ-geometry of the space S as the study of the properties of the space and its configurations that are invariant under [the group] ξ. That is, ξ-geometry of the space S is the study of the ξ-geometry of S. (Wilder [38], p. 180.)

The concept of geometry *à la Klein* can, of course, be formulated in set theory (see Klein [24] and [25]). Let D be a set and g a group of transformations of D. A geometry in the sense of Klein is a structure $G = \langle\langle D, r_\iota\rangle g\rangle$, such that g is the group of all transformations of D which leave relations r_ι invariant, and any invariant relation in $\varepsilon(D)$ is expressible in $\langle D, r_\iota\rangle$ in terms of r_ι. We call 'figure' any element of $\varepsilon(D)$ defined by a formula of the language of G, i.e., that of set theory plus D and r_ι. Two figures A and B are said to be equipollent if they are of the same type and there is a transformation of g that transforms A onto B. Equipollence is an equivalence relation on the set of figures of a fixed type. The invariants of figures are said to be invariants of G.

In most cases, D is already endowed with some structure, and the relations of this structure are employed to define the relations r_ι. If g is the group of automorphisms of $e = \langle D, s_\lambda\rangle$, then $e' = \langle\langle D, s_\lambda\rangle, g\rangle$ is a Klein geometry. It is clear, then, that the concept of geometry underlies all mathematics. (An important generalization of Klein's geometry is obtained by means of the notion of the action of a group on a structure, but here is not the place to pursue this issue.)

Another conception of geometry is that of Blumenthal and Menger [3], to which we now turn. Any figure F of $G = \langle\langle D, r_\iota\rangle g\rangle$ belongs to $\varepsilon(D)$ and so can be set-theoretically constructed starting with some subset K of D. The structure $\langle K, r_\iota^*, F\rangle$, where r_ι^* is the restriction of r_ι to K, restriction easily definable, is called the structure associated with F. Given two figures F_1 and F_2, we can prove the following propositions:

PROPOSITION 1. *If F_1 and F_2 are equipollent, then their associated structures are isomorphic (as substructures of $\langle D, r_\iota\rangle$). (We show below that de converse of this first proposition is not true.)*

PROPOSITION 2. *Any substructure of $\langle D, r_\iota\rangle$ is a figure.*

Two figures are said to be BM-equipollent if their associated structures are isomorphic. Any BM-equipollence constitutes a relation of equivalence in the class of all figures of a given type. For Blumenthal and Menger, a geometry over a structure $e = \langle D, r_\iota\rangle$ is a pair $H = \langle e, \approx_t\rangle$, where $\approx_t$ is a BM-equipollence between figures of type t. Loosely speaking, in the study of H we are interested in properties of figures that are invariant under BM-equipollences.

The connections between Klein geometries (K-geometries) and those of Blumenthal-Menger (BM-geometries) are clear: every K-geometry is a BM-geometry, but, as will become clear below, there are BM-geometries that are not K-geometries.

In the definition of BM-geometry, the structure $\langle e, \approx_t\rangle$ may be such that e is trivial, *i.e.*, a bare set. Blumenthal and Menger note that:

> This widens the applicability of the notion. We may, for example, consider the theory of cardinal numbers as a geometry over a set Σ, by defining two figures (subsets) of Σ to be equivalent if and only if there exists a one-to-one correspondence between their elements. But in most of the important geometries, the convention that establishes the equivalence of figures makes use of the *structure* by virtue of which a set becomes the element-set of a space [structure]. Hence, one speaks more often of a geometry over a space than over a set. (Blumenthal and Menger [3], p. 28.)

Clearly, this remark is also true of K-geometry.

In Riemannian geometry and some of its extensions, due to the fact that the corresponding automorphism groups are commonly trivial, composed by the identity transformation only, Klein's methodology isn't enough to specify satisfactorily a geometry as a K-geometry. It is not difficult to realize that the inclusive view of Blumenthal and Menger cannot be profitably applied either. To address this problem, Veblen devised an interesting approach (see Veblen and Whitehead [37]).

According to Veblen, a pseudo-group is a non-empty set S of bijective transformations of subsets of a fixed set, such that: (1) The inverse of any transformation of S belongs to S. (2) If there is the product of two transformations of S, then it is in S. Obviously, every transformation group is a pseudo-group.

Let us now suppose, in what follows, that all functions satisfy standard conditions of continuity and differentiability. Veblen then defines, in a differential manifold M, a geometric object k as follows: (a) k is defined at every point P of M; (b) k determines, in every coordinate system S, at P, an ordered set of m real numbers, called the components of k in S at P (m may be different from the dimension of M); (c) the components of k in any coordinate system S', at P, are functions of the components of k in S at P, the functions representing the coordinate transformations and the derivatives of these functions evaluated at P.

As a result, geometric objects are defined in every point of the manifold and are characterized by their transformation laws. They are, in fact, invariant under the pseudo-group of point transformations induced by local coordinate transformations. Examples of these objects are: (1) scalars, *i.e.*, objects that have only one component, which is invariant under coordinate transformations, (2) vectors, and (3) tensors.

A differential geometry, in the sense of Veblen, is a triple $V = \langle H, g, k \rangle$, in which M is a differential manifold, g is a pseudo-group as above, and k is a geometric object. When k is a scalar, V is essentially the manifold M. When k is a (metric) tensor, V is a Riemannian geometry. (See Kobayashi and Nomizu [26] for a discussion of differential geometry.)

The BM-metric geometry associated to a 'space' with a distance function defined between any two of its points, following Blumenthal and Menger, is the BM-geometry in which BM-equipollence is the relation of isomorphism between figures of a fixed type.

Here are some examples: (1) Geometries according to Klein: (1.1) the topology of $\mathcal{R}^3$ is the geometry whose group is the group the homeomorphisms of $\mathcal{R}^3$; (1.2) the 'logic' of a set is the K-geometry whose group of transformations is composed by all transformations of the set; (1.3) the affine geometry of $\mathcal{R}^3$, in the present context, corresponds to the group of affinities; (1.4) the Euclidean metric geometry is characterized by the group of isometries of $\mathcal{R}^3$. (2) BM-geometries: (2.1) the metric geometry of an infinitely dimensional Hilbert space; (2.2) the analogous metric geometry of a Minkowski space (the space of special relativity). (3) Veblen geometries (V-geometries): (3.1) any Riemannian geometry is a V-geometry; (3.2) differential manifolds are V-geometries, in which the geometric object is a scalar.

Geometries were envisaged above as set-theoretic structures. However, they can also be considered as species of structures or, more in line with current mathematical practice, as certain kind of theories. The axiomatic approach to geometry is firmly based on language, as is evident. In various cases, this language is that of set theory extended by new primitive terms (and postulates), but typed languages can also be employed, for instance those associated with the structures under study. Even when a particular geometry is formulated in a higher-order framework, it is possible, as we have already observed, to investigate its lower parts. This happens in the cases of the Euclidean real plane and of arithmetic (another example: we could investigate first-order versions of special relativity). (Axiomatic geometry, however, will not be taken into consideration here, since it is not relevant to the objectives of this paper.)

Let us return to structures, and let $t = \langle E, T \rangle$ be a topological space inside ZF, in which E is the base space and T its topology. The language in which we talk about t is, thus, that of ZF plus the constants t, E and T; the postulates are those of ZF plus the common topological axioms with some extra ones (such as those of Hausdorff). We denote by G the group of homeomorphisms (the automorphisms) of t. Clearly any topological notion, that is, an element of the scale of t, definable in the structure in the strict sense or in the wide sense, is invariant under G. But an important point is also the converse question: Is any invariant notion definable in function of the primitive notions of t? In this connection, we have:

THEOREM 3. *In t, any invariant notion d (that is, an element of the scale of t) is definable in the wide sense in the infinitary language associated with t whenever d is invariant under G.*
Proof. *To say that d is definable in the wide sense in t means that d is expressible in the primitive constants of t. So, the theorem is a consequence of the results of da Costa and Rodrigues [16].*

THEOREM 4. *In Euclidean metric geometry, a notion is definable in the wide sense (that is, expressible) if, and only if, it is invariant under the group of isometries.*
Proof. *See da Costa and Rodrigues [16].*

In general, propositions analogous to the preceding ones are valid for all structures.

THEOREM 5. *There exist BM-geometries that aren't K-geometries.*
Proof. *It suffices to observe that the BM-metric geometry of an infinite-dimensional Hilbert space is the study of isometries between figures of the space, and that there are*

isomorphisms between two figures that cannot be extended to an isometry of the whole space onto itself.

THEOREM 6. *There are structures in which some notions are expressible (definable in the wide sense) but not definable in the strict sense in the denumerable language associated with the structure.*

Proof. *In ZF the set of real numbers $\mathcal{R}$ is rigid: $\mathcal{R}$ has only one automorphism, its identity isomorphism. As a result, according to da Costa and Rodrigues [16], any notion of the scale of $\mathcal{R}$ is expressible, i.e., definable in the wide sense, in the infinitary language associated with $\mathcal{R}$. However, since $\mathcal{R}$ is not denumerable, there are real numbers that are not definable in the (finitary) language of ZF plus the primitive symbols corresponding to primitive notions of $\mathcal{R}$.*

We also have:

THEOREM 7. *Any structure can be extended to a rigid structure by the addition of a finite number of new primitive relations.*

Proof. *The proof of this theorem of Sebastião e Silva is presented in da Costa and Rodrigues [16].*

The Galois group of a Riemannian geometry V is usually trivial. So, in this case, any notion (element of the scale of V) is only definable in the wide sense in V. Moreover, there are notions that are not definable in the strict sense in V, with the help of the underlying language of ZF, as is clear.

5 Indiscernibility and related concepts

Since the beginning of quantum theory, in the early part of the twentieth century, notions such as indiscernibility and identity have been repeatedly discussed in the context of physics. For instance, Schrödinger argued, several times, that the notion of identity is meaningless in connection with elementary particles (see French and Krause [20] for a thorough discussion of this issue).

We now consider the analysis of the notions of indiscernibility, identity, and individuality in the domain of geometry. We take this to be a first step toward a future analysis of these concepts in the context of quantum physics. A key motivation here is the work done by Caulton and Butterfield [11].

Recall that our discussion is carried out in set theory, in which languages are both set-theoretic constructs and geometric structures. Geometry has, of course, two main faces, since there is a parallelism between geometry as a mathematical discipline and geometry as a physical theory underlying other physical theories. Both can be formulated in set-theoretically.

Identity, informally, constitutes the connection between objects a and b which are the same. We write, in this case, $a = b$. In set theory, 'identity' has two meanings: it may denote a basic concept of set theory or it may refer to the diagonal of the Cartesian product of a non-empty set by itself, i.e., a set-theoretic relation, formed by a set of ordered pairs whose two coordinates are the same. It is clear that, in set theory, infinitely many *relations* of identity can be expressed. In what follows, the reader will easily recognize the meaning in which term 'identity' is being employed.

Our set-theoretic ZF system is a pure set theory, in the sense that the only objects taken into consideration are sets. Identity, in the two set-theoretic senses above, can be defined or can be taken as a primitive notion. (On whether identity in general — understood as a fundamental concept rather than as a notion in set theory — can in fact be defined is an issue explored in Bueno [8] and [9].)

Geometric objects belong to scales of structures. So identity commonly refers to objects in certain geometric structures that are associated with languages. To talk about identity, in this case, is normally to talk about relations in the scales of structures, associated with abstract languages.

Identity is independent of language: if a and b are, for instance, figures in the scale of structure S, then that a is identical to b (or is different from b) constitutes a relation that is independent of the language associated with S. But this is not true of the notion of indiscernibility, which generalizes the concept of identity. We distinguish three kinds of indiscernibility: the weak, the strong, and the intrinsic kinds. They extended concepts presented in Caulton and Butterfield [11], although we use a different terminology.

5.1 Weak indiscernibility

By $e = \langle D, r_\iota \rangle$ we denote, in the next definitions, a first-order structure in which r_ι is a finite sequence, and L_e is the first-order finitary language associated with e, and x and y are two distinct variables of L_e. We then have the following definitions, in which r stands for any element of the sequence r_ι.

DEFINITION 8.

1. If r is a unary predicate the formula:

   $$r(x) \leftrightarrow r(y)$$

   is called the companion formula of r.

2. If r is a binary relation, then the formula:

   $$\forall z[(r(x,z) \leftrightarrow r(y,z)) \wedge (r(z,x) \leftrightarrow r(z,y))]$$

   is the companion formula of r.

3. If r is a ternary relation, then

   $$\forall z \forall t[(r(x,t,z) \leftrightarrow r(y,t,z)) \wedge (r(t,x,z) \leftrightarrow r(t,y,z)) \wedge (r(t,z,x) \leftrightarrow r(t,z,y))]$$

   is the companion formula of r.

4. The same goes for any n-ary relation r, with n finite and greater than three.

Note that new variables $t, z, \ldots$ are different from x and y, and any two of them are distinct.

DEFINITION 9. $x \equiv y$ abbreviates the conjunction of all companions of the elements of r_ι.

DEFINITION 10. If a and b are numbers of the domain of $e = \langle D, r_\iota \rangle$, we say that they are weakly indiscernible (*wi*) if $a \equiv b$ is true in e.

Definitions 1, 2 and 3 can be modified to accommodate first-order languages and structures with infinitely many primitive relations, as well as infinitary languages. We now consider higher-order languages and structures.[4]

DEFINITION 11. If $e = \langle D, r_\iota \rangle$ is a higher-order structure with one of its associated languages, and a and b are elements of the scale of e, then:

1. If a and b have the same type and if there are companions of all or some of the relations r_ι, then $a \equiv b$ is similarly defined as in the case of first-order languages and structures. Under these assumptions, a is weakly indiscernible from b if, and only if, $a \equiv b$ is true in $\varepsilon(D)$; otherwise, $x \equiv y$ is $x = x \wedge y = y$ and a and b are also said to be weakly indiscernible.

2. If a and b aren't weakly indiscernible, then a and b are weakly discernible.

Here are some examples: (a) $\mathcal{R}^3$, with the standard metric d, constitutes a rigid structure whose Galois group is composed by the identity transformation. The same goes for the Klein geometry $K = \langle \langle \mathcal{R}^3, d \rangle, g \rangle$, in which g is its group. As a result, in the associated infinitary languages, any two distinct points of $\mathcal{R}$ or of K are weakly discernible. The same point applies to two distinct spheres. (b) Any two points of an abstract projective plane are indiscernible (see, for instance, Heyting [22]).

5.2 Strong (or syntactic) indiscernibility

Let a and b be two figures of the same type (in particular, points) of the geometry à la Klein $e = \langle \langle D, r_\iota \rangle, g \rangle$. In this case, a and b are strongly discernible if there is a formula $F(x)$, of the language associated with e, which has x as its sole free variable, such that, in $\varepsilon(D)$, it is true that:

$$\models F(a) \text{ and } \models \neg F(b).$$

If there is no such formula, a and b are said to be strongly indiscernible. It is clear that this definition can be extended to any structure, and depends on the language associated with it.

As for examples, we have: (1) In the usual higher-order, metric geometry of $\mathcal{R}^3$, two spheres, whose centers are strongly indiscernible, are also strongly indiscernible. (The language of the usual geometry is finitary and, as a consequence, there are real numbers that are strongly indiscernible.) (2) In the usual axiomatizations of Euclidean geometry, for example the one carried out by Hilbert, any two points are always strongly indiscernible.

---

[4]On higher-order model theory and infinitary model theory, which have connections with the general theory of structures, see, for example, Manin [30] and the references therein. Shelah insists that, in higher-order model theory, the central feature is not the language but certain algebraic properties of families of structures. In particular, it is not possible to study properly such structures in first-order languages.

5.3 Essential (or semantic) indiscernibility

In the Klein geometry $G = \langle\langle D, r_\iota\rangle, g\rangle$, any two figures a and b are said to be essentially indiscernible if there is a transformation of g that maps a onto b. Obviously, this definition can be extended to any structures whatsoever. When two figures of the scale of a structure are not essentially indiscernible, they are essentially discernible.

Here are a couple of examples: (1) In the metric geometry of $\mathcal{R}^3$, figures of the same type with domains essentially indiscernible are essentially indiscernible. (2) In the usual propositional calculus, considered as a free Boolean algebra, any two generators are essentially indiscernible.

Two other kinds of indiscernibility (or of discernibility), investigated by Caulton and Butterfield [11], can be extended to higher-order languages or to infinitary ones. However, we will not examine them here. We only note that Caulton and Butterfield's absolute discernibility corresponds, in first-order languages, to our syntactic discernibility.

There are connections among our three concepts of discernibility (and indiscernibility) that hold for all mathematical structures. Using ideas of da Costa and Rodrigues [16], the following theorems can be proved:

THEOREM 12. *In any structure $e = \langle D, r_\iota\rangle$, with the language L_e, if two objects of e are syntactically discernible, then they are semantically discernible.*

THEOREM 13. *The converse of theorem 5.1 is not true in general.*

THEOREM 14. *Under the conditions of Theorem 5.1, if L_e is the strict infinitary language associated with e (see Section 1 above), then syntactic discernibility is equivalent to semantic discernibility.*

Let $e = \langle D, r_\iota\rangle$ be a structure and L_e its language. An individual (or an individual of type i) is an element of D that is syntactically indiscernible from all other elements of D. More generally, an individual of type t is an element of type t that is syntactically indiscernible from all other elements of type t.

In some cases, the objects of type t, $t \neq i$, can be the basic elements of the domain of a new structure and, in this new structure, they are individuals (of type i of the new structure). It is clear that the notion of indiscernibility depends not only on the structure we are considering, but also on the language employed.

In first-order languages, weak indiscernibility can be used as a kind of identity (see, for instance, Caulton and Butterfield [11]), although clearly it isn't identity *per se*. The situation in higher-order structures and languages is entirely different. In this case, weak indiscernibility is too restrictive to have a similar role. In effect, the primitive objects of $\varepsilon(D)$ that correspond to the structure $e = \langle D, r_\iota\rangle$ don't have the same order and type, and r_ι belongs to various types, so that weak indiscernibility becomes vacuous. Perhaps we could introduce weak indiscernibility for each type. However, it is better to use identity. But in the case of first-order structures or first-order languages, weak indiscernibility is relevant.

Most of the discussion in Caulton and Butterfield [11] that is devoted to first-order structures and languages can be extended to the higher-order case.

For instance, in elementary Euclidean geometry, that is, Hilbert geometry, two spheres with the same radius are syntactically indiscernible. However, in the metric geometry of $\mathcal{R}^3$, any two spheres with radii of identical length are syntactically discernible when their centers are also syntactically discernible.

With regard to Caulton and Butterfield's metaphysical views, their discussion can also be extended to the higher-order situation. To fix our ideas, we mention that their considerations on discernibility of pairs of objects are immediately generalizable to higher-order objects. In fact, several of their metaphysical views are adaptable to higher-order situations. In other words, their first-order metaphysics is extendable to a higher-order one. As an illustration, there can be higher-order individuals, similarly to the first-order individuals defined by them.

6 Structural metaphysics

In this section we outline a metaphysics of structures. Our preliminary analysis is intuitive and informal, but we will indicate how the approach can be developed rigorously and formally. Our work is not exegetical, and we are not attempting to systematize the views defended by contemporary structural metaphysicians (see, for instance, Ladyman [29], French and Ladyman [19], and French [18]). Our aim is to present a possible metaphysics of structures inspired by these views.

Consider the notion of mathematical structure:

$$e = \langle D, r_\iota \rangle$$

as introduced above. A particular case of structure is, obviously, that in which the sequence of primitive relations is empty:

$$e = \langle D, \emptyset \rangle.$$

When $D = \emptyset$, we have the empty structure that was not discussed above. When r_ι is $\emptyset$, the structure e can be considered as the set D. So, from an informal stance, it seems reasonable to regard sets as a kind of structure.

Moreover, any relation is a strict relation, that is, an n-ary relation with $n \geq 2$, or a unary relation or set. Furthermore, relations are sets or, what is the same thing, unary relations. In particular, a strict n-ary relation, $n \geq 2$, is also a unary relation, that is, a set.

The structure:

$$e = \langle D, r_\iota \rangle$$

is, basically, the binary relation:

$$\{\langle D, r_\iota \rangle\}.$$

Conversely, an n-adic relation r, $n \geq 2$, essentially constitutes the structure:

$$r^* = \langle K, r \rangle,$$

where K is a set $\bigcup \{\{x\} : x \in r\}$.

As a result of these informal considerations, sets, relations and set-theoretic structures are, as it were, different faces of the same kind of basic objects: metaphysical structures or, to simplify, M-structures. Certain M-structures are fundamental: they are the M-sets, the bricks that compose all M-structures.

We can postulate, then, that M-sets constitute a system satisfying the axioms of a traditional set theory, such as ZFC, Zermelo-Fraenkel set theory with the axiom of choice (see Fraenkel and Bar Hillel [17], and Kuratowski and Mostowski [28]). However, the postulates of the theory should not include the axiom of regularity, given our assumption that any M-structure can be formed by other structures. In this way, our system reduces to ZF.

In addition to these considerations, it is natural to take ZF as providing a systematization of a metaphysics of M-structures, independently of its standard interpretation as a theory of usual sets or collections. However, we need to accommodate M-structures that are invoked in the representation of the physical universe (or, in a metaphysical reading, in the constitution of that universe). For this reason, we need to accommodate space and time. Accordingly, we add to ZF a new unary predicate $\circ$, according to which $\circ(x)$ means that x is a physical (or concrete) M-structure or M-Urelement (for example, particles may be conceived of as unary M-relations, or M-structures, etc.) The resulting system will be denoted by ZF°. Postulates governing M-structures that satisfy $\circ$ are added to those of ZF, depending on the view one has on the nature of the universe. We are also free to strengthen ZF° in various ways, for instance in order to handle M-categories and M-toposes interpreted as M-structures. (On ZF with *Urelemente*, see Brignole and da Costa [5].)

We can now formulate standard scientific theories in an ontology of M-structures in analogy with the use of common set-theoretic structures in the foundations of science. In particular, fields and particles can be interpreted in terms of different kinds of M-structures. The philosophical simplicity that an ontology of M-structures brings to the philosophy of science, especially of physics, should be apparent.

Note that although ZF° was formulated in analogy with set-theoretic ideas, it does possess other interpretations, even non set-theoretic ones. This is the case of the pure M-structures, which can be used to capture some features of structuralist approaches in the philosophy of mathematics and science.

M-structures can also be used to formulate Hughes' suggestive idea of structural explanation. As he notes:

> The idea of a structural explanation can usefully be approached via an example of special relativity. Suppose we were asked to explain why one particular velocity (in fact the speed of light) is invariant across the set of inertial frames. The answer offered in the last decade of the nineteenth century was to say that measuring rods shrank at high speeds in such way that a measurement of this velocity in a moving frame always gave the same value as one in a stationary frame. This causal explanation is now seen as seriously misleading; a much better answer would involve sketching the models of space-time which special relativity provides and showing that in these models, for a certain family of pairs

of events, not only is their spatial separation x proportional to their temporal separation t, but that the quantity x/t is invariant across admissible (that is, inertial) coordinate systems; further, for all such pairs, x/t always has the same value. This answer makes no appeal to causality; rather, it points out structural features of the models that special relativity provides. It is, in fact, a structural explanation. (Hughes [23], pp. 256-257.)

M-structures, suitably developed, provide a straightforward way of formulating the relevant models of special relativity and the invariances across various coordinate systems.

One need not, however, reify these invariant features of the models: they indicate traits that, according to the models under consideration, have a significant form of objectivity. It is not up to us that these invariant features emerge: they are stable properties of the relevant models. Of course, realists will read these traits as salient features of the world, whereas anti-realists will resist such interpretation. What matters to us is the understanding that emerges from each interpretation and the fact that both can be captured in terms of M-structures. Explanation, including structural explanation, is clearly tied to understanding, and M-structures provide a rich framework to explore all of these alternatives. (We return to this point below.)

Realist views of structures insist that structures either provide us with ways of knowing the world (epistemic conceptions) or articulate ways the world is (ontic views). (For the distinction between epistemic and ontic views of structural realism, see Ladyman [29]; see also French [18] for additional discussion. For a critique of ontological conceptions of structure, see Arenhart and Bueno [1].) In contrast, anti-realist views of structure resist such commitments, and insist that structures, despite their help in systematizing information about the world, should not be reified. Due to its partiality, incompleteness, imprecision or inaccuracy, the information encoded in such structures is at best partial (for a thorough defense of the significance of partiality in this context, see da Costa and French [15]; see also Bueno [6]).

The proposal here is best thought of as being neutral between either side of the debate: it need not be committed to either realism about structures or to anti-realism about them. Structures can be used in geometry, physics or metaphysics quite independently of their particular interpretation, quite independently of the particular framework in which they are formulated. Set theory can be employed to formulate a K-geometry without stating that this geometry metaphysically is a particular set-theoretic structure. One can simply use set-theoretic resources as *representational tools* rather than as *constitutive devices*. That is, set theory provides a framework to represent and formulate various structures, but the framework itself need not be taken to answer the metaphysical question of the nature of the objects and structures that are represented in it. Nor is the framework indispensable. One can, after all, use different frameworks to carry out the task at hand: for instance, in addition to the many non-equivalent formulations of set theory, category theory, type theory, second-order mereology, among other frameworks, can all be employed as well. And in each of these frameworks, importantly different answers are given to the metaphysical question of the nature of the relevant structures: categorical, type-theoretic, mereological, etc. Since the metaphysical nature of the objects and structures is different in different frameworks, and all of these frameworks are possible, it's unclear that any particular metaphysical answer is ultimately settled.

Even *M*-structures, despite being decidedly formulated in set theory, need not be taken to be *metaphysically* set-theoretic objects, since there are corresponding counterparts to such structures in all of the different frameworks mentioned above. On this interpretation, set theory is used as a representational tool only without any indication of the fundamental nature of the structures that are thus represented.

This distinction between set theory as a tool of representation and as a constitutive device can be used in other contexts in which set theory is employed. As is well known, Tarski often highlighted that he was a nominalist (for references and discussion, see Frost-Arnold [21]). Given all the work he did in set theory as well as his use of it in semantics and theories of truth, if set theory were ontologically committed to sets as abstract objects (as the usual platonist reading would have it), Tarski would end up being incoherent. However, if set theory is simply interpreted as a representational tool, which need not capture or state the fundamental nature of the objects under study, one can use it while still resisting commitment to the existence of the objects and structures it refers to. (In addition, neutral quantifiers, which do not presuppose the existence of the objects that are quantified over, can also be invoked. Thus quantifying over sets is not enough to be ontologically committed to their existence: additional conditions of an existence predicate also need to be met; for details, see Azzouni [2] and Bueno [7].) This has the advantage of making Tarski's stance coherent, while providing an alternative setting to understand the use of structures in a variety of domains, from geometry through physics to metaphysics.

We noted above the significance of structural explanations (and theoretical explanations) in physics. Are there such explanations in metaphysics? Clearly there should be. Certainly there are no causal explanations in this area, but still explanatory devices are systematically invoked. A metaphysical problem can be thought of as something that typically emerges from incompatible situations in (our conceptual description of) the world. Metaphysical explanations are then attempts to resolve the (apparent) tension. These explanations are considerations that indicate the possibility of certain situations in light of circumstances that seem to undermine them. For instance, how are individuals possible at the fundamental level given some interpretations of non-relativist quantum mechanics that deny that identity can be applied to quantum particles? How can abstract (that is, causally inert, non-spatial-temporal) mathematical structures exist given a world constituted by concrete (causally active, spatial-temporal) entities? Metaphysical explanations indicate that the apparent tension between these circumstances can be resolved. In some cases they suggest ways of reconciling the conflicting situations; in others, they provide reasons to undermine some of the built-in assumptions invoked in the description of the circumstances under consideration; in yet others, they suggest ways in which the inconsistency identified among the relevant situations can be tolerated.

In each case, some understanding emerges: we may realize the consistency of situations we thought were actually impossible; we may identify assumptions for which we may lack good reason to accept; we may learn about a broader range of possibilities (which may eventually include what was presumed to be, up to that point, impossibilities). In each case, our understanding increases, by gaining information about ways of reconciling what was thought to be incompatible, by obtaining information regarding false or questionable assumptions, or by increasing the scope of what is deemed possible. Metaphysical explanations are, thus, crucial for our understanding of several aspects of our conceptual framework. (For the general structure of philosophical explanations as ways of dispelling apparent tensions between conflicting assumptions, see Nozick [31].)

We think that M-structures provide a rich framework in which we can examine and assess these metaphysical explanations as they emerge in the foundations of the sciences. We can clearly identify the relevant assumptions: set-theoretic, category-theoretic, type-theoretic, mereological, etc., depending on the context, as well as empirical, theoretical and conceptual presuppositions. We also probe each framework's expressive and inferential power and its limitations. As a result, a proper assessment of their pros and cons can be determined and, in this way, some progress is eventually made. And our understanding, in the end, increases.

BIBLIOGRAPHY

[1] Arenhart, J., and Bueno, O., Structural Realism and the Nature of Structure. *European Journal for Philosophy of Science* **5**: 111-139, 2015.
[2] Azzouni, J., *Deflating Existential Consequence*. New York: Oxford University Press, 2004.
[3] Blumenthal, L. M. and Menger, K., *Studies in Geometry*. San Francisco: W. H. Freenman and Company, 1970.
[4] Bourbaki, N., *Theory of Sets*. London-Ontario: Hermann-Addison-Wesley, 1968.
[5] Brignoli, D. and da Costa, N. C. A., On Supernormal Ehresmann-Dedecker Universes. *Mathematische Zeitschrift* 122: 342-350, 1971.
[6] Bueno, O., Empirical Adequacy: A Partial Structures Approach. *Studies in History and Philosophy of Science* 28: 585-610, 1997.
[7] Bueno, O., Dirac and the Dispensability of Mathematics. *Studies in History and Philosophy of Modern Physics* 36: 465-490, 2005.
[8] Bueno, O., Why Identity is Fundamental. *American Philosophical Quarterly* 51: 325-332, 2014.
[9] Bueno, O., Can Identity Be Relativized?. In: Koslow and Buchsbaum (eds.) (2015), 253-262.
[10] Corry, L., *Modern Algebra and the Rise of Mathematical Structures*. Berlin: Birkhäuser Verlag, 1996.
[11] Caulton, A. and Butterfield, J., On Kinds of Indiscernibility in Logic and Metaphysics. *British Journal for the Philosophy of Science* 63: 27-84, 2012.
[12] da Costa, N. C. A., Some Aspects of Quantum Physics. *Principia* 11: 77-95, 2007.
[13] da Costa, N. C. A. and Bueno, O., Remarks on Abstract Galois Theory. *Manuscrito* 34: 151-183, 2011.
[14] da Costa, N. C. A. and Chuaqui, R., On Suppes' Set-Theoretical Predicates. *Erkenntnis* 29: 95-112, 1988.
[15] da Costa, N. C. A. and French, S., *Science and Partial Truth*. New York: Oxford University Press, 2003.
[16] da Costa, N. C. A. and Rodrigues, A. A. M., Definability and Invariance. *Studia Logica* 86: 1-30, 2007.
[17] Fraenkel, A. A. and Bar-Hillel, Y., *Foundations of Set theory*. Amsterdam: North-Holland, 1958.
[18] French, S., *The Structure of the World*. Oxford: Oxford University Press, 2014.

[19] French, S. and Ladyman, J., Remodelling Structural Realism: Quantum Physics and the Metaphysics of Structure. *Synthese* 136: 31-56, 2003.
[20] French, S. and Krause, D., *Identity in Physics*. Oxford: Oxford University Press, 2006.
[21] Frost-Arnold, G., Tarski's Nominalism. In: Patterson (ed.) (2008), 225-246.
[22] Heyting, A., *Axiomatic Projective Geometry*. Amsterdam: North-Holland, 1988.
[23] Hughes, R.I.G., *The Structure and Interpretation of Quantum Mechanics*. Cambridge, MA: Harvard University Press.
[24] Klein, F., Vergleichende Betrachtungen über neuere geometrische Forschungen. *Mathematische Annalen* 43: 63-100, 1893.
[25] Klein, F., *Le Programme d'Erlangen*. Paris-Montréal: Gauthier-Villars, 1974.
[26] Kobayashi, S. and Nomizu, K., *Foundations of Differential Geometry, Volume I*. New York-Toronto: John Willey, 1963.
[27] Koslow, A. and Buchsbaum, A. (eds.) *The Road to Universal Logic, volume II*. Dordrecht: Birkhäuser, 2015.
[28] Kuratowski, K. and Mostowski, A., *Set Theory*. Amsterdam: North-Holland, 1968.
[29] Ladyman, J., What is structural realism? *Studies in History and Philosophy of Science* 29: 409-424, 1998.
[30] Manin, Yu. I., *A Course in Mathematical Logic for Mathematicians*. New York: Springer, 2010.
[31] Nozick, R., *Philosophical Explanations*. Cambridge, MA: Harvard University Press, 1981.
[32] Patterson, D. (ed.) *New Essays on Tarski and Philosophy*. Oxford: Oxford University Press, 2008.
[33] Shoenfield, J. R., *Mathematical Logic*. Reading-London. Addison-Wesley, 1967.
[34] Suppes, P., *Representation and Invariance of Scientific Structures*. Stanford: CSLI Publications, 2002.
[35] Tarski, A., *A Decision Method for Elementary Algebra and Geometry*. Berkeley: University of California Press, 1951.
[36] Veblen, V. and Young, J. W., *Projective Geometry. (2 volumes.)* Boston-London: Ginn and Company, 1938/1946.
[37] Veblen, V. and Whitehead, J. H. C., *The Foundations of Differential Geometry*. Cambridge: Cambridge University Press, 1953.
[38] Wilder, R. L., *Introduction to the Foundations of Mathematics*. New York-London: John Willey, 1960.

What Algebraic Properties of Quantities Are Needed to Model Accelerated Observers in Relativity Theory

GERGELY SZÉKELY

ABSTRACT. We investigate the possible structures of physical quantities over which accelerated observers can be modeled in special relativity. We present a general axiomatic theory of accelerated observers which has a model over every real closed field. We also show that, if we would like to model certain accelerated observers, then not every real closed field is suitable, e.g., uniformly accelerated observers cannot be modeled over the field of real algebraic numbers. Consequently, the class of fields over which uniform acceleration can be investigated is not axiomatizable in the language of ordered fields.

1 Introduction

In this paper, within an axiomatic framework, we investigate the possible structures of physical quantities over which accelerated observers can be modeled in special relativity.

There are several reasons for this kind of investigations. One of them is that we cannot experimentally verify whether the structure of quantities is isomorphic to $\mathbb{R}$ (the field of real numbers). Thus we cannot have any direct empirical support for leaving out of consideration other algebraic structures that offer themselves naturally. Moreover, the outcomes of physical experiments are always finite decimals (hence rational numbers). Despite this fact almost all physical theories assume that the structure of quantities is $\mathbb{R}$.

These investigations lead to a deeper understanding of the relation between our mathematical and physical assumptions. For a more general perspective of this research direction, see Andréka *et. al.* (2012b).

In general, we would like to investigate the question

"What structure can quantities have in a certain physical theory?"

To introduce our central concept, let Th be an axiomatic theory of physics in which the structure of quantities can be defined. In this case, we can introduce notation $Num(\text{Th})$ for the class of the possible structures of quantities over which theory Th can be modeled:

$$Num(\text{Th}) := \{\mathfrak{Q} : \mathfrak{Q} \text{ is a structure of quantities over which Th has a model.}\}$$

In this paper, our main question of interest is what algebraic properties the physical quantities have to satisfy if we want to model accelerated observers in special relativity. So here we restrict our investigation to the case when Th is a theory of special relativity extended with accelerated observers. However, this question can be investigated in any other physical theory the same way.

Throughout this paper, we restrict our investigation to spacetimes which are at least 3-dimensional (i.e., spacetimes in which space is at least 2-dimensional) because 2-dimensional spacetimes behave differently even from the point of view of our main investigation, see Andréka et. al. (2012b).

In Section 2, we recall several theories and axioms for relativity theory from the literature. For example, $\mathsf{SpecRel_d}$ (see p.164) is an axiom system for d-dimensional special relativity that captures the kinematics of special relativity since it implies that the worldview transformations between inertial observers are Poincaré transformations, see Andréka et. al. (2012b).

It is easy to see that $\mathsf{SpecRel_d}$ has a model over every ordered field, i.e.,

$$Num(\mathsf{SpecRel_d}) = \{\mathfrak{Q} : \mathfrak{Q} \text{ is an ordered field}\}.$$

In particular, $\mathsf{SpecRel_d}$ has a model over the field $\mathbb{Q}$ of rational numbers, too. However, if we add to $\mathsf{SpecRel_d}$ the assumption that inertial observes can move with arbitrary speed less than that of light, then every positive quantity has to have a square root, see Andréka et. al. (2012b). In particular, the structure of quantities cannot be the field $\mathbb{Q}$ of rational numbers, but it can be the field of real algebraic numbers.

If we require only that observers can move with *approximately* any speed slower than that of light, then we still can model special relativity over the field $\mathbb{Q}$ of rational numbers, see Madarász and Székely (2013).

We also recall our theory $\mathsf{AccRel_d}$ of accelerated observers from Székely (2009), see p.6. $\mathsf{AccRel_d}$ requires the structure of quantities to be a real closed field, i.e., an ordered field in which every positive quantity has a square root and every odd degree polynomial has a root. Specially, $\mathsf{AccRel_d}$ does not have a model over $\mathbb{Q}$. However, any real closed field can be the structure of quantities of $\mathsf{AccRel_d}$, i.e.,

$$Num(\mathsf{AccRel_d}) = \{\mathfrak{Q} : \mathfrak{Q} \text{ is a real closed field}\},$$

see Theorem 1 at p.165. In particular, $\mathsf{AccRel_d}$ can be modeled over the field of real algebraic numbers.

The main result of this paper is that if we extend $\mathsf{AccRel_d}$ by an extra axiom stating that there are uniformly accelerated observers ($\mathsf{Ax\exists UnifOb}$, see p.166), then the field of real algebraic numbers cannot be the structure of quantities anymore, see Theorem 2. A surprising consequence of this result is that $Num(\mathsf{AccRel_d} + \mathsf{Ax\exists UnifOb})$ is not a first-order logic axiomatizable class of fields, see Corollary 3. That is, in the language of ordered fields, it is impossible to axiomatize those fields over which uniformly accelerated observers can be modeled.

In a related approach, Stannett introduces two structures for the quantities belonging to relativity theory: one for the measurable quantities and one for the theoretical ones, see Stannett (2011).

2 Axiomatic Framework

Here we use the following two-sorted language of first-order logic parametrized by a natural number $d \geq 3$ representing the dimension of spacetime:

$$\{B, Q; \mathsf{Ob}, \mathsf{IOb}, \mathsf{Ph}, +, \cdot, \leq, \mathsf{W}\}, \tag{1}$$

where B (bodies) and Q (quantities) are the two sorts, Ob (observers), IOb (inertial observers) and Ph (light signals) are one-place relation symbols of sort B, $+$ and $\cdot$ are two-place function symbols of sort Q, $\leq$ is a two-place relation symbol of sort Q, and W (the worldview relation) is a $d+2$-place relation symbol the first two arguments of which are of sort B and the rest are of sort Q.

Relations $\mathsf{Ob}(o)$, $\mathsf{IOb}(m)$ and $\mathsf{Ph}(p)$ are translated as "*o is an observer,*" "*m is an inertial observer,*" and "*p is a light signal,*" respectively. To speak about coordinatization, we translate relation $\mathsf{W}(k, b, x_1, x_2, \ldots, x_d)$ as "*body k coordinatizes body b at space-time location $\langle x_1, x_2, \ldots, x_d \rangle$,*" (i.e., at instant x_1 and space location $\langle x_2, \ldots, x_d \rangle$).

Formulas are built up from these basic concepts in the usual way. For the precise definition of the syntax and semantics of first-order logic, see, e.g., Chang and Keisler (1990) §1.3, §2.1, §2.2.

We use the notation Q^n for the set of all n-tuples of elements of Q. If $\bar{x} \in Q^n$, we assume that $\bar{x} = \langle x_1, \ldots, x_n \rangle$, i.e., x_i denotes the i-th component of the n-tuple $\bar{x}$. Specially, we write $\mathsf{W}(m, b, \bar{x})$ in place of $\mathsf{W}(m, b, x_1, \ldots, x_d)$, and we write $\forall \bar{x}$ in place of $\forall x_1 \ldots \forall x_d$, etc.

To make them easier to read, we omit the outermost universal quantifiers from the formalizations of our axioms, i.e., all the free variables are universally quantified.

The key axiom of special relativity states that the speed of light is the same in every direction for every inertial observers.

> **AxPh**: For any inertial observer, the speed of light is the same everywhere and in every direction (and it is finite). Furthermore, it is possible to send out a light signal in any direction (existing according to the coordinate system) everywhere:[1]
>
> $$\mathsf{IOb}(m) \to \exists c_m \Big[c_m > 0 \land \forall \bar{x}\bar{y} \Big(\exists p \big[\mathsf{Ph}(p) \land \mathsf{W}(m, p, \bar{x}) \land \mathsf{W}(m, p, \bar{y}) \big]$$
> $$\leftrightarrow (x_2 - y_2)^2 + \ldots + (x_d - y_d)^2 = c_m^2 \cdot (x_1 - y_1)^2 \Big) \Big].$$

To get back the intended meaning of axiom AxPh (or even to be able to define subtraction from addition), we have to assume some properties of quantities.

In our next axiom, we state some basic properties of addition, multiplication and ordering true for real numbers.

---

[1] That is, if m is an inertial observer, then there is a positive quantity c_m such that for all coordinate points $\bar{x}$ and $\bar{y}$ there is a light signal p coordinatized at $\bar{x}$ and $\bar{y}$ by observer m if and only if equation $(x_2 - y_2)^2 + \ldots + (x_d - y_d)^2 = c_m^2 \cdot (x_1 - y_1)^2$ holds.

AxOField: The structure of quantities $\langle Q, +, \cdot, \leq \rangle$ is an ordered field, i.e.,

- $\langle Q, +, \cdot \rangle$ is a field in the sense of abstract algebra; and
- the relation $\leq$ is a linear ordering on Q such that
    i) $x \leq y \to x + z \leq y + z$ and
    ii) $0 \leq x \wedge 0 \leq y \to 0 \leq xy$ holds.

Using axiom AxOFiled instead of assuming that the structure of quantities is the field of real numbers not just makes our theory more flexible, but also makes it possible to meaningfully investigate the main question of this paper.

Another reason for using AxOField instead of $\mathbb{R}$ is that we cannot experimentally verify whether the structure of physical quantities are isomorphic to $\mathbb{R}$. The two axioms of real numbers which are the most difficult to defend from empirical point of view are the Archimedean axiom, see Rosinger (2008), Rosinger (2009) §3.1, Rosinger (2011b), Rosinger (2011a), and the supremum axiom, see the remark after the introduction of CONT on p.165.

We also support AxPh with the assumption that all observers coordinatize the same "external" reality (the same set of events):

AxEv All inertial observers coordinatize the same set of events.

For formalization of axioms AxEv, as well as that of AxSelf and AxSymD (below), see, e.g., Andréka *et. al.* (2012a).

The three axioms above are enough to capture the essence of special relativity. However, we usually assume the following two simplifying axioms.

AxSelf: Inertial observers are stationary relative to themselves.

AxSymD: Any two inertial observers agree as to the spatial distance between two events if these two events are simultaneous for both of them; and the speed of light is 1 for all observers.

Axiom AxSymD is a symmetry axiom saying that inertial observers use the same units of measurement.

Let us introduce an axiom system for special relativity as the collection of the five axioms above:

$$\mathsf{SpecRel_d} := \mathsf{AxPh} + \mathsf{AxOField} + \mathsf{AxEv} + \mathsf{AxSelf} + \mathsf{AxSymD}.$$

Let us also recall our first-order logic axiom system of accelerated observers $\mathsf{AccRel_d}$. The key axiom of $\mathsf{AccRel_d}$ is the following:

AxCmv: At each moment of its worldline, each observer sees the nearby world for a short while as an inertial observer does.

For formalization of axiom AxCmv, as well as AxEv$^-$ and AxSelf$^-$ (below), see, e.g., Andréka *et. al.* (2012a), Székely (2009).

In $\mathsf{AccRel_d}$ we also use the following localized version of axioms AxEv and AxSelf of $\mathsf{SpecRel_d}$.

AxEv$^-$: Observers coordinatize all the events in which they participate.

AxSelf$^-$: In his own worldview, the worldline of any observer is a subinterval of the time-axis.

Let us now introduce a promising theory of accelerated observers as SpecRel$_d$ extended with the three axioms above.

$$\text{AccRel}_d^0 := \text{SpecRel}_d + \text{AxCmv} + \text{AxEv}^- + \text{AxSelf}^-.$$

Axiom AxCmv ties the behavior of accelerated observers to that of the inertial ones, and SpecRel$_d$ captures the kinematics of special relativity. Therefore, it is quite natural to hope that AccRel$_d^0$ is a strong enough theory of accelerated observers to prove the most fundamental results about accelerated observers. However, AccRel$_d^0$ does not imply even the most basic predictions about accelerated observers such as the twin paradox or that stationary observers measure the same time between two events Madarász, Németi and Székely (2006), Székely (2009) §7. Moreover, it can be proved that even adding the whole first-order logic theory of real numbers to AccRel$_d^0$ is not enough for getting a theory that implies the twin paradox, see Madarász, Németi and Székely (2006), Székely (2009) §7.

In the models of AccRel$_d^0$ in which TwP is not true, there are some definable gaps in $\mathfrak{Q}$. Our axiom schema CONT below excludes these gaps.

CONT: Every parametrically definable, bounded and nonempty subset of Q has a supremum (i.e., least upper bound) with respect to $\leq$.

In CONT "definable" means "definable in the language of AccRel$_d$." For formulation of CONT, see Madarász, Németi and Székely (2006)p. 692 or Székely (2009) §10.1.

Axiom schema CONT makes the supremum postulate of real numbers closer to the physical/empirical level because CONT speaks only about "physically meaningful" subsets of the quantities which can be defined in the language of our (physical) theory and not about "any fancy subset."

When $\mathfrak{Q}$ is the ordered field of real numbers, CONT is automatically true. Our axiom system AccRel$_d$ is the extension of AccRel$_d^0$ by axiom schema CONT.

$$\text{AccRel}_d := \text{AccRel}_d^0 + \text{CONT}.$$

It can be proved that axiom system AccRel$_d$ implies the twin paradox, see Madarász, Németi and Székely (2006), Székely (2009) §7.2.

An ordered field is called **real closed field** if its every positive element has a square root, and every polynomial of odd degree has a root in it. It is known that a field is real closed iff it is elementarily equivalent to the field $\mathbb{R}$ of real numbers (i.e., they are indistinguishable in the language of ordered fields by first-order logic formulas), see Tarski (1951).

Axiom schema CONT is so powerful that it implies that the possible structures of quantities have to be real closed fields:

THEOREM 1.

$$Num(\text{AccRel}_d) = \{\mathfrak{Q} : \mathfrak{Q} \text{ is real closed fields}\}.$$

Theorem 1 is proved in Székely (2009), see Proposition 10.1.2 there.

QUESTION 1. Can CONT be replaced in AccRel$_d$ with some natural assumptions such that they (together with AccRel$_d^0$) imply all (or certain) important predictions of relativity theory about accelerated observers (e.g., the twin paradox) yet they do not require that the structure of quantities is a real closed field?

3 Structure of quantities required by uniformly accelerated observers

In Andréka et. al. (2012b), we have seen that assuming existence of observers can ensure the existence of certain quantities. So let us investigate another axiom of this kind, which postulates the existence of uniformly accelerated observers. To introduce this axiom, let us define the **life-curve** lc$_{km}$ of observer k according to observer m as the worldline of k according to m *parametrized by the time measured by k*, formally:

$$\mathsf{lc}_{km} := \{\langle t, \bar{x}\rangle : \exists \bar{y} \ k \in \mathsf{ev}_k(\bar{y}) = \mathsf{ev}_m(\bar{x}) \wedge y_1 = t\}, \tag{2}$$

where ev$_m(\bar{x})$, the event occurring for observer m at point $\bar{x}$, is defined as the set of bodies m coordinatizes at $\bar{x}$:

$$\mathsf{ev}_m(\bar{x}) := \{b : \mathsf{W}(m, b, \bar{x})\}. \tag{3}$$

Now we can introduce our axiom ensuring the existence of uniformly accelerated observers. It can be proved that uniformly accelerated observers are moving along hyperbolas in relativity theory, see, e.g., d'Inverno (1992) §3.8, pp.37-38. For the sake of simplicity, we use this fact in the formalization of the following axiom:

Ax∃UnifOb: It is possible to accelerate an observer uniformly:

$$\mathsf{IOb}(m) \to \exists k \Big[\mathsf{Ob}(k) \wedge \mathsf{Dom}\,\mathsf{lc}_{km} = Q \wedge \mathsf{lc}_{km}(0) = \bar{y} \wedge$$
$$\mathsf{lc}_{km}(1)_1 > y_1 \wedge \forall \bar{x}\big[\bar{x} \in \mathsf{Ran}\,\mathsf{lc}_{km} \leftrightarrow (x_2 - y_2)^2 - (x_1 - y_1)^2 = a^2 \wedge$$
$$x_3 - y_3 = \ldots = x_d - y_d = 0\big]\Big].$$

We use the notation $\mathfrak{Q} \in Num(\mathsf{Th})$ for algebraic structure $\mathfrak{Q}$ the same way as the model theoretic notation $\mathfrak{Q} \in Mod(\mathsf{AxField})$, e.g., $\mathbb{Q} \in Num(\mathsf{Th})$ means that $\mathbb{Q}$, the field of rational numbers, can be the structure of quantities in Th.

Let $\mathbb{A} \cap \mathbb{R}$ denote the ordered field of real algebraic numbers. Theorem 2 states that the ordered field of real algebraic numbers cannot be the structure of quantities of theory AccRel$_d$ + Ax∃UnifOb:

THEOREM 2.
$$\mathbb{A} \cap \mathbb{R} \notin Num(\mathsf{AccRel}_d + \mathsf{Ax\exists UnifOb}).$$

The detailed proof of Theorem 2 is in Section 4. The key idea of the proof is that a kind of exponential function can be defined over all the structures of quantities over which $\mathsf{AccRel_d + Ax\exists UnifOb}$ can be modeled (see Theorem 8), but this kind of exponential function cannot be defined over the field of real algebraic numbers.

Since the ordered fields of real numbers and real algebraic numbers are elementarily equivalent, Theorem 2 implies that the structure of quantities of $\mathsf{AccRel_d + Ax\exists UnifOb}$ is not an elementary class:[2]

COROLLARY 3. *The class* $Num(\mathsf{AccRel_d + Ax\exists UnifOb})$ *is not axiomatizable in the language of ordered fields.*

By Theorem 2, we know that not every real closed field can be the structure of quantities in $\mathsf{AccRel_d + Ax\exists UnifOb}$, e.g., it cannot be the field of real algebraic numbers. However, the following questions are still open:

QUESTION 2. *Exactly which fields are in* $Num(\mathsf{AccRel_d + Ax\exists UnifOb})$?

Analogously, we can also ask what properties of quantities do axiom $\mathsf{Ax\exists UnifOb}$ requires without CONT:

QUESTION 3. *Exactly which fields are in* $Num(\mathsf{AccRel_d^0 + Ax\exists UnifOb})$?

Theorem 8 on p.170 suggests that the answer to Questions 2 and 3 may have something to do with ordered exponential fields, see Dahn and Wolter (1983) §4, Kuhlmann (2000).

4 Proof of Theorem 2

To prove Theorem 2, let us introduce some concepts. The **space component** of $\bar{x} \in Q^d$ is defined as

$$\bar{x}_s := \langle x_2, \ldots, x_d \rangle. \tag{4}$$

The (signed) **Minkowski length** of $\bar{x} \in Q^d$ is

$$\mu(\bar{x}) := \begin{cases} \sqrt{x_1^2 - |\bar{x}_s|^2} & \text{if } x_1^2 \geq |\bar{x}_s|^2, \\ -\sqrt{|\bar{x}_s|^2 - x_1^2} & \text{in other cases,} \end{cases} \tag{5}$$

and the **Minkowski distance** between $\bar{x}$ and $\bar{y}$ is $\mu(\bar{x}, \bar{y}) := \mu(\bar{x} - \bar{y})$. We use the signed version of the Minkowski length because it contains two kinds of information: (i) the length of $\bar{x}$, and (ii) whether it is spacelike, lightlike or timelike. Let $H \subseteq Q$. We say that H is an **interval** iff $z \in H$ when there are $x, y \in H$ such that $x < z < y$. We say that a function $\gamma : H \to Q^d$ is a **curve** if H is an interval and has at least two distinct elements.

The usual (first-order logic) formula can be used to define the differentiability function over any ordered field $\mathfrak{Q}$. The **derivative of** function $f : Q \to Q^n$ is $A \in Q^n$ at $x_0 \in \text{Dom } f$:

$$\text{Diff}(f, x_0, A) \overset{\text{def}}{\iff} \forall \varepsilon > 0 \, \exists \delta > 0 \, \forall x \; 0 < |x - x_0| < \delta$$
$$\wedge \; x \in \text{Dom } f \to |f(x) - f(x_0) - A(x - x_0)| < \varepsilon |x - x_0|. \tag{6}$$

---

[2] Of course, it is a pseudo elementary class; it is the class of ordered field reducts of elementary class $Th(\mathsf{AccRel_d + Ax\exists UnifOb})$.

In the case when there is one and only one A such that $\text{Diff}(f, x_0, A)$ holds, we write $f'(x_0) = A$. It can be proved that there is at most one A such that $\text{Diff}(f, x_0, A)$ holds if $\text{Dom } f$ is open, see Székely (2009) *Thm.10.3.9*.

A curve γ is called **timelike curve** iff it is differentiable, and $\gamma'(t)$ is timelike, i.e., $\mu(\gamma'(t)) > 0$, for all $t \in \text{Dom } \gamma$. We call a timelike curve α **well-parametrized** if $\mu(\alpha'(t)) = 1$ for all $t \in \text{Dom } \alpha$.

THEOREM 4. *Assume* AccRel_d. *Let k be an observer and m be an inertial observer. Then* lc_{km} *is a well-parametrized timelike curve. See Székely (2009) Thm.6.1.11.*

A part of real analysis can be generalized for arbitrary ordered fields without any real difficulty, see Székely (2009) §10. However, a certain fragment of real analysis can only be generalized within first-order logic for *definable* functions and their proofs need axiom schema CONT. We refer to these generalizations by marking them "CONT-." The first-order logic generalizations of some theorems, such as Chain Rule can be proved without CONT, so they are naturally referred to without the "CONT-" mark.

PROPOSITION 5. *Assume* CONT *and* AxOField. *Let $\gamma, \delta : Q \to Q^d$ be definable and differentiable well-parametrized timelike curves such that* $\text{Ran } \gamma = \text{Ran } \delta$. *Then there are $\varepsilon \in \{-1, +1\}$ and $c \in Q$ such that $\delta(t) = \gamma(\varepsilon t + c)$ for all $t \in Q$.*

Proof. By Székely (2009) *Lem.10.5.4*, we have that there is a (definable) differentiable function $h : Q \to Q$ such that $|h'| = 1$ and $\delta(t) = \gamma(h(t))$ for all $t \in Q$. By CONT-Darboux's Theorem Székely (2009) p.110, $h'(t) = 1$ for all $t \in Q$ or $h'(t) = -1$ for all $t \in Q$. By the CONT version of the fundamental theorem of integration Székely (2009) *Prop.10.3.19*, $h(t) = t + c$ or $h(t) = -t + c$ for some $c \in Q$. ∎

LEMMA 6. *Assume* AccRel_d. *Let m_1, m_2 be inertial observers and let k_1 and k_2 be observers such that*

1. $\text{Dom lc}_{k_1 m_1} = \text{Dom lc}_{k_2 m_2} = Q$,
2. $\text{Ran lc}_{m_1 k_1} = \text{Ran lc}_{m_2 k_2}$,
3. $\text{lc}_{k_1 m_1}(0) = \text{lc}_{k_2 m_2}(0)$,
4. $\text{lc}_{k_1 m_1}(1)_1 > \text{lc}_{k_1 m_1}(0)_1$ *and* $\text{w}_{k_2 m_2}(1)_1 > \text{w}_{k_1 m_1}(0)_1$.

Then $\text{lc}_{k_1 m_1} = \text{lc}_{k_2 m_2}$.

Proof. We are going to prove our statement by applying Proposition 5 to $\text{lc}_{k_1 m_1}$ and $\text{lc}_{k_2 m_2}$. So let $\gamma := \text{lc}_{k_1 m_1}$ and $\delta := \text{lc}_{k_2 m_2}$. By Theorem 4, γ and δ are well-parametrized timelike curves. By assumptions (1) and (2) $\text{Dom } \gamma = \text{Dom } \delta = Q$ and $\text{Ran } \gamma = \text{Ran } \delta$. Therefore, by Proposition 5, there is a $c \in Q$ and $\varepsilon \in \{-1, +1\}$ such that $\delta(t) = \gamma(\varepsilon t + c)$ for all $t \in Q$. By assumption (3), $\gamma(0) = \delta(0)$. Therefore, $c = 0$. Since γ and δ are timelike curves γ_1 and δ_1 are either strictly increasing or strictly decreasing functions. By assumption (4), $\gamma(1)_1 > \gamma(0)_1$ and $\delta(1)_1 > \delta(0)_1$. Thus both γ_1 and δ_1 are strictly increasing. Consequently, $\gamma'(0)_1 > 0$ and $\delta'(0)_1 > 0$. Therefore, ε cannot be negative. Hence we have that $\varepsilon = 1$. Consequently, $\gamma = \delta$ as it was stated. ∎

THEOREM 7. *Assume* AccRel$_d$ *and* Ax∃UnifOb. *There are definable differentiable functions* $S : Q \to Q$ *and* $C : Q \to Q$ *with the following properties:*

1. $C^2 - S^2 = 1$,
2. $S(0) = 0$ *and* $C(0) = 1$,
3. $S(1) > 0$,
4. $C(-t) = C(t)$ *and* $S(-t) = S(t)$ *for all* $t \in Q$,
5. $(S')^2 - (C')^2 = 1$,
6. $C' = S$ *and* $S' = C$,
7. S *is strictly increasing on* Q; *and* C *are strictly increasing on interval* $[0, \infty)$ *and strictly decreasing on* $(-\infty, 0]$.
8. Ran $S = Q$ *and* Ran $C = [1, \infty)$.

Proof. Let binary relation on observers H be defined as

$$H(m,k) \stackrel{def}{\iff} \mathsf{IOb}(m) \wedge \mathsf{Ob}(k)$$
$$\wedge \text{ Dom } \mathsf{lc}_{km}(t) = Q \wedge \mathsf{lc}_{km}(0) = \langle 0,1,0\ldots 0\rangle \wedge \mathsf{lc}_{km}(1)_1 > 0$$
$$\wedge \forall \bar{x} [\bar{x} \in \text{Ran } \mathsf{lc}_{km} \leftrightarrow x_2^2 - x_1^2 = 1 \wedge x_3 = \ldots = x_d = 0]. \quad (7)$$

Let γ be defined as the following relation:

$$\gamma(t) = \bar{x} \stackrel{def}{\iff} \forall mk \ H(m,k) \wedge \mathsf{lc}_{km}(t) = \bar{x}. \quad (8)$$

By axiom Ax∃UnifOb, there are such observers k and m that $H(m,k)$ holds. Therefore relation γ is not empty. By Lemma 6, γ is a function and it equals to lc_{km} for any observers k and m for which relation $H(m,k)$ holds.

Dom $\gamma = Q$ by Dom $\mathsf{lc}_{km}(t) = Q$. By Theorem 4, γ is a well-parametrized timelike curve.

Let $C = \gamma_2$ and $S = \gamma_1$. Then $C : Q \to Q$ and $S : Q \to Q$ are definable differentiable functions since they are coordinate functions of definable differentiable function $\gamma(t) = \langle S(t), C(t), 0 \ldots 0\rangle$.

Item (1) holds since Ran $\gamma = \{\bar{x} : x_2^2 - x_1^2 = 1 \wedge x_3 = \ldots = x_d = 0\}$, Item (2) holds since $\gamma(0) = \langle 0,1,0\ldots 0\rangle$, and Item (3) holds since $\gamma_1(1) > 0$ by the definition of $H(m,k)$. Item (5) holds since $\gamma' = \langle S', C', 0, \ldots, 0\rangle$ because γ is well-parametrized.

To prove Item (4), let us consider curve $\delta : t \mapsto \langle -S(t), C(t), 0 \ldots, 0\rangle$. It is clear that Dom δ = Dom γ. By Item (1), Ran δ = Ran γ. It is clear that δ is also a well-parametrized curve by Item (5). Therefore, by Proposition 5, $\delta(t) = \gamma(\varepsilon t + c)$ for all $t \in Q$. By Item (1), $\delta(0) = \gamma(0)$. Thus $c = 0$. By Chain Rule, $\delta'(t) = \varepsilon \gamma'(\varepsilon t)$. Since both δ and γ are well-parametrized curves, $\delta(0) = \gamma(0) = \langle 0,1,0,\ldots,0\rangle$, and the tangent line of Hyperbola $\{\bar{x} : x_1^2 - x_1^2 = 1, x_3 = \ldots = x_d\}$ is vertical, we have that $\gamma_1'(0) = 0$ and $\gamma_2'(0) = 0$. Thus

$\delta(0) = \langle -1, 0, \ldots, 0 \rangle$. Hence $\varepsilon = -1$. Thus $\delta(t) = \gamma(-t)$. Consequently, $-S(t) = S(-t)$ and $C(t) = C(-t)$.

By Székely (2009) Thm.10.2.4, S and C are monotonous on interval $[0,s]$ for all $0 < s \in Q$. Hence they are also monotonous on interval $[0, \infty)$. By Item (5), $(\gamma_1')^2 \geq 1 > 0$. Therefore, by CONT-Darboux Theorem, see Székely (2009) §10.3, $S'(t) > 0$ for all $t \in Q$ or $S'(t) < 0$ for all $t \in Q$. $\gamma(1)_1 > \gamma(0)_1$ by Items (2) and (3). Therefore, γ_t is increasing. Thus $\gamma_t' > 0$. So S is strictly increasing on Q. Item (1), C is strictly increasing on $[0, \infty)$ and strictly decreasing on $(-\infty, 0]$ since S strictly increasing on Q.

Now let us prove Item (6). We have $S'^2 - C'^2 = 1$ by Item (6). By Chain Rule, if we differentiate both sides of this equation, we get that $2SS' - 2CC' = 0$. Hence $C'C = S'S$. Multiplying $S'^2 - C'^2 = 1$ by S^2, we get $S'^2 S^2 - C'^2 S^2 = S^2$. From this we get $C'^2(C^2 - S^2) = S^2$ by $C'C = S'S$. Therefore, $C'^2 = S^2$ since $C^2 - S^2 = 1$. Consequently, $C'(t) = \pm S(t)$ and $S'(t) = \pm C(t)$ for all $t \in Q$. By Items (1) and (7), $S'(t) > 0$ and $C(t) > 0$. Therefore, $S' = C$. If $t > 0$, a similar argument show that $C'(t) = S(t)$. By (4), $-C'(t) = C'(-t)$ and $S(-t) = -S(t)$. Therefore, $C'(t) = S(t)$ also if $t \in (-\infty, 0]$. By Item (1) and (5), $C'(0) = 0$ and $S'(0) = 1$. Hence $C'(t) = S(t)$ for all $t \in Q$.

By CONT-Boltzano Theorem Székely (2009) §10.2, Ran S and Ran C are intervals. Therefore Item (8) holds by Item 1. ∎

THEOREM 8. *Assume* $\mathsf{AccRel_d}$ *and* $\mathsf{Ax\exists UnifOb}$. *There is a definable differentiable function* $E : Q \to Q$ *with the following properties:*

1. $E(0) = 1$,
2. $E(1) > 0$,
3. $E(-t)E(t) = 1$,
4. $E' = E$,
5. Ran $E = (0, \infty)$ *and*
6. E *is strictly increasing.*

Let the **restriction** of function f to set H be defined as

$$f|_H := \{\langle x, y \rangle : x \in \text{Dom } f \cap H \text{ and } y = f(x)\} \tag{9}$$

Proof. Let $S : Q \to Q$ and $C : Q \to Q$ be the definable differentiable functions which exist by Theorem 7. Let $E := C + S$. Then E is a definable differentiable function since C and S are so. Items (1) and (2) follow directly from Items (2) and (3) of Theorem 7. Item (3) follows from Items (1) and (4) of Theorem 7 because $E(-t)E(t) = (C(-t) + S(-t))(C(t) + S(t)) = C^2(t) - S^2(t) = 1$. Item (5) follows from Item (6) of Theorem 7. Item (4) follows because of the following. Ran $E|_{[0,\infty)} = [1, \infty)$ because $E(0) = 1$, S and C are strictly increasing on $[0, \infty)$, and Ran $C = [1, \infty)$ by Item (8) of Theorem 7. Hence Ran $E|_{(-\infty,0]} = (0, 1]$ by Item (3). Thus Ran $E = (0, \infty)$. Item (6) follows from Item (6) of Theorem 7 since E is strictly increasing on $[0, \infty)$ by Item (7) of Theorem 7 and E is also strictly increasing on $(-\infty, 0]$ since $E(-t)E(t) = 1$ by Item (3). ∎

The following first-order logic formula defines that **limit of function** f is A at x_0 over every ordered field:

$$\text{Limit}(f, x_0, A) \overset{def}{\iff} \forall \varepsilon > 0 \; \exists \delta > 0 \; \forall x$$
$$0 < |x - x_0| < \delta \wedge x \in \text{Dom } f \to |f(x) - A| < \varepsilon. \quad (10)$$

In the case when there is one and only one A such that $\text{Limit}(f, x_0, A)$ holds, we write $\lim_{x \to x_0} f(x) = A$. By using the technique of Székely (2009) §10, it can be proved that there is at most one A such that $\text{Limit}(f, x_0, A)$ holds if x_0 is a accumulation point of Dom f (i.e., if for all $\varepsilon > 0$, there is $x \in \text{Dom } f$ such that $|x - x_0| < \varepsilon$).

Let the exponential function of $\mathbb{R}$ be denoted by exp.

PROPOSITION 9. *Let $\langle Q, +, \cdot, \leq \rangle$ be a subfield of $\mathbb{R}$.[3] Let $f : Q \to Q$ be a differentiable function such that $f' = f$ and $f(0) = 1$. Then $f = \exp|_Q$, i.e., f is the restriction of the real exponential function to Q.*

Proof. We have that f is continuous since f is differentiable, see, e.g., Székely (2009) Cor.10.3.5. Let $f_*(x) := \lim_{t \to x} f(t)$. Function f_* is well defined since f is continuous and Q is dense in $\mathbb{R}$ (as $Q \subseteq \overline{Q}$). Since f is continuous, we also have that f_* is an extension of f, i.e., $f_*(x) = f(x)$ if $x \in Q$. We are going to show that $f_* = \exp$. First we show that $f'_*(x) = f_*(x)$. We start by showing that, for all $x, y \in \mathbb{R}$ and $\varepsilon_0 > 0$, there are $x^*, y^* \in Q$ such that $|x - x^*| < \varepsilon_0$, $|y - y^*| < \varepsilon_0$, and

$$\left| \frac{f(x^*) - f(y^*)}{x^* - y^*} - \frac{f_*(x) - f_*(y)}{x - y} \right| < \varepsilon_0.$$

By the triangle inequality,

$$\left| \frac{f_*(x) - f_*(y)}{x - y} - \frac{f(x^*) - f(y^*)}{x^* - y^*} \right| \leq$$
$$\left| \frac{f_*(x) - f_*(y)}{x - y} - \frac{f(x^*) - f(y^*)}{x - y} + \frac{f(x^*) - f(y^*)}{x - y} - \frac{x - y}{x^* - y^*} \cdot \frac{f(x^*) - f(y^*)}{x - y} \right|$$
$$\leq \left| \frac{f_*(x) - f(x^*)}{x - y} \right| + + \left| \frac{f_*(y) - f(y^*)}{x - y} \right| + \left| 1 - \frac{x - y}{x^* - y^*} \right| \cdot \left| \frac{f(x^*) - f(y^*)}{x - y} \right|. \quad (11)$$

By the definition of f_*, there is a δ such that

$$|f_*(x) - f(x^*)| < \frac{\varepsilon |x - y|}{3} \quad \text{and} \quad |f_*(y) - f(y^*)| < \frac{\varepsilon |x - y|}{3} \quad (12)$$

if $|x - y| < \delta$. From this, by the triangle inequality, we have that

$$|f(x^*) - f(y^*)| \leq |f(x^*) - f_*(x)| + |f_*(x) - f_*(y)|$$
$$+ |f_*(y) - f(y^*)| < |f_*(x) - f_*(y)| + \frac{2\varepsilon_0 |y - x|}{3}. \quad (13)$$

---

[3] By Pickert–Hion Theorem, these fields are exactly the fields of Acrhiedean ordered fields, see, e.g., Fuchs (1963) §VIII, Mikhalev and Pilz (2002) C.44.2.

Since $\left|1 - \frac{x-y}{x^*-y^*}\right|$ can be arbitrarily small if $|x - x^*|$ and $|y - y^*|$ are small enough, we can choose x^* and y^* such that

$$\left|1 - \frac{x-y}{x^*-y^*}\right| \cdot \left|\frac{f(x^*) - f(y^*)}{x-y}\right| < \frac{\varepsilon_0}{3}. \tag{14}$$

Therefore, there are x^* and y^* arbitrarily close to x and y such that

$$\left|\frac{f_*(x) - f_*(y)}{x-y} - \frac{f(x^*) - f(y^*)}{x^*-y^*}\right| < \varepsilon_0. \tag{15}$$

To prove that $f'_* = f_*$, We have to show that, for all $\varepsilon > 0$, there is a $\delta > 0$ such that

$$\left|f_*(x) - \frac{f_*(x) - f_*(y)}{x-y}\right| < \varepsilon \tag{16}$$

if $|x - y| < \delta$. By the triangle inequality,

$$\left|f_*(x) - \frac{f_*(x) - f_*(y)}{x-y}\right| \leq |f_*(x) - f(x^*)|$$
$$+ \left|f(x^*) - \frac{f(x^*) - f(y^*)}{x^*-y^*}\right| + \left|\frac{f(x^*) - f(y^*)}{x^*-y^*} - \frac{f_*(x) - f_*(y)}{x-y}\right|. \tag{17}$$

By the definition of f_*,

$$|f_*(x) - f(x^*)| < \frac{\varepsilon}{3} \tag{18}$$

if $|x - x^*|$ small enough. By (15), we have that there, are x^* and y^* arbitrarily close to x and y such that,

$$\left|\frac{f(x^*) - f(y^*)}{x^*-y^*} - \frac{f_*(x) - f_*(y)}{x-y}\right| < \frac{\varepsilon}{3}. \tag{19}$$

Since $f' = f$, we have that

$$\left|f(x^*) - \frac{f(x^*) - f(y^*)}{x^*-y^*}\right| < \frac{\varepsilon}{3} \tag{20}$$

if $|x^* - y^*|$ is small enough. So, if $|x - y|$ is small enough and we can choose x^* and y^* close enough to x and y we have that Eq. (20) holds. Consequently, if $|x - y|$ is small enough, then Eq. (16) holds, i.e., f_* is differentiable and $f'_* = f_*$. Therefore, there is a $c \in \mathbb{R}$ such that $f_*(x) = c \exp(x)$ for all $x \in \mathbb{R}$. We have that $c = 1$ since $c = c \exp(0) = f_*(0) = f(0) = 1$. Therefore, f is the restriction of function $\exp$ to Q; and this is what we wanted to prove. ∎

Proof.[Proof of Theorem 2] By Theorem 8, a differentiable function E is definable in the models of $\mathsf{AccRel_d} + \mathsf{Ax\exists UnifOb}$ such that $E' = E$ and $E(0) = 1$. By Proposition 9, E has to be the restriction of $\exp$ to the real algebraic numbers. However, this is impossible since then $E(1)$ is the Euler-number e which is not an algebraic number. ∎

Acknowledgements

I am grateful to Hajnal Andréka for her valuable comments and suggestions for improving the quality of this paper. This research is supported by the Hungarian Scientific Research Fund for basic research grants No. T81188 and No. PD84093.

BIBLIOGRAPHY

[1] H. Andréka, J. X. Madarász, I. Németi, and G. Székely. A logic road from special relativity to general relativity. *Synthese*, 186(3):633–649, 2012a.
[2] H. Andréka, J. X. Madarász, I. Németi, and G. Székely. What are the numbers in which space-time?, 2012b. arXiv:1204.1350.
[3] C. C. Chang and H. J. Keisler. *Model theory*. North-Holland Publishing Co., Amsterdam, 1990.
[4] B. I. Dahn and H. Wolter. On the theory of exponential fields. *Z. Math. Logik Grundlag. Math.*, 29(5):465–480, 1983.
[5] R. d'Inverno. *Introducing Einstein's relativity*. Oxford University Press, New York, 1992.
[6] H. B. Enderton. *A mathematical introduction to logic*. Academic Press, New York, 1972.
[7] L. Fuchs. *Partially ordered algebraic systems*. Pergamon Press, Oxford, 1963.
[8] S. Kuhlmann. *Ordered exponential fields*, volume 12 of *Fields Institute Monographs*. American Mathematical Society, Providence, RI, 2000.
[9] J. X. Madarász, I. Németi, and G. Székely. Twin paradox and the logical foundation of relativity theory. *Found. Phys.*, 36(5):681–714, 2006.
[10] Judit X. Madarász and Gergely Székely. Special relativity over the field of rational numbers. *International Journal of Theoretical Physics*, pages 1–13, 2013.
[11] Alexander V. Mikhalev and Günter F. Pilz, editors. *The concise handbook of algebra*. Kluwer Academic Publishers, Dordrecht, 2002.
[12] E. E. Rosinger. Two essays on the archimedean versus non-archimedean debate, 2008. arXiv:0809.4509.
[13] E. E. Rosinger. Special relativity in reduced power algebras, 2009. arXiv:0903.0296.
[14] E. E. Rosinger. Cosmic contact to be, or not to be archimedean. *Prespacetime Journal*, 2(2):234–248, 2011a.
[15] E. E. Rosinger. How far should the principle of relativity go? *Prespacetime Journal*, 2(2):249–264, 2011b.
[16] M. Stannett. Computing the appearance of physical reality. *Appl. Math. Comput.*, in press, 2011.
[17] G. Székely. *First-Order Logic Investigation of Relativity Theory with an Emphasis on Accelerated Observers*. PhD thesis, Eötvös Loránd Univ., Budapest, 2009.
[18] A. Tarski. *A decision method for elementary algebra and geometry*. University of California Press, Berkeley and Los Angeles, Calif., 1951.

Laws, Accidental Generalities, and the Lotze Uniformity Condition

ARNOLD KOSLOW

ABSTRACT. There has been a great deal of discussion of what laws are, but very little about accidental generalizations. We develop a new account of the difference between these generalizations that depends upon a reconstruction of an insight of the German philosopher Hermann Lotze, and uses the notion of explanation to provide conditions for being a law and being an accidental generalization rather than the other way round. Some elementary consequences follow with the aid of certain closure principles for explanation.

Keywords: Accidental generalizations; Explanatory closure; Laws; Lotze uniformity condition.

1 The Lotze Uniformity Condition

It has become a hallmark of any satisfactory account of scientific laws, that they cannot be accidental generalizations. There is of course a very easy way of assuring the difference, and that is by opting for a definition of accidental generalizations as exactly those that are not lawlike. That option is something of a cheat. Philosophers who invoke the accidental feature of a generalization usually do so to explain why a true generalization is not a law. Thus, Hempel says that the generalization "All the rocks in that box contain iron" fails to be a law because it is accidental that those rocks all share that property. Obviously Hempel isn't just reminding us of a definition; he's giving a reason.[1] The follow up question then is this: why is the rocks in the box generalization accidental? Something more is needed.

On this point J. Carroll has astutely noted that most of the current discussions have focused on scientific laws and not much can be found on accidental generalizations.[2]

Our proposal for marking this distinction among generalizations, has an interesting source. One immediate source is W.E. Johnson who distinguished two different kinds of generalized conditional: universals of fact, and universals of law:

---

[1] Prentice Hall, 1966, p.55.
[2] "Nailed to Hume's Cross" in *Contemporary Debates in Metaphysics*, eds. J. Hawthorne and D. Zimmerman, Blackwell, 2008.

> It will have been observed that the correlative notions of determination and dependence enter into the formulation of the principles as directly applicable to the characters of manifestations and therefore only derivatively to the manifestations themselves. Hence the potential range for which these principles hold extends beyond the actually existent into the domain of the possibly existent. In this way the universality of law is wider than that of fact. While the universals of fact are implied by universals of law, the statement of the latter has intrinsic significance not involved in that of the former.[3] (*Logic*, Part I, 1921, pp.251-252).

This is a remarkable observation of Johnson. He says that there are two kinds of uniformity (something which Mill, he notes, failed to separate). He says that the universals of fact are applicable to objects (manifestations), and the universals of law apply directly to the properties (characters) of those objects. Moreover, he required that universal conditionals follow from laws. Finally he required that all the universals of law should be stated as what we now call counterfactual conditionals.

> Thus taking two determinate adjectives p and q under the respective determinables P and Q, the factual universal may be expressed in the form 'Every substantive PQ in the universe of reality is q if p; while the assertion of law assumes the form 'Any substantive PQ in the universe of reality would be q if it were p.' These formulae represent fairly, I think, the distinction which Mill had in mind; for my first formula may be said to express a mere invariability in the association of q with p. while the second expresses the unconditional connection between q and p. Or as I have said in p.252, Chapter XIV, Part I, the universal of fact covers only the actual, whereas the universal of law extends beyond the actual into the range of the possible.[4]

Although Johnson believed that Mill had such a distinction in mind, we think that another more likely source for the contrast between two kinds of conditionals, can be found in the logical writings of the nineteenth century German philosopher Hermann Lotze, with whose work on Logic Johnson was acquainted. It is not immediately evident from Johnson's writings that he had read Lotze. He is nowhere cited by Johnson. Nevertheless the evidence for this is clear since Johnson was lecturing on Lotze's *Logik* in 1888, in his Cambridge course on logic. At that time, in Cambridge, lecturers listed the syllabus for their course of lectures on a printed formal card, a brochure of a sort, which was presumably made available to students. There is in the Cambridge University Archives, a volume of the minutes of the Board of Research stud-

---

[3] *Logic Part I*, Cambridge University Press 1921, pp. 251-252. Reprinted by Dover Publications, N.Y. 1964.

[4] *Logic Part III, The Logical Foundations of Science*, Cambridge University Press, 1924. P.6. Reprinted by Dover Publications NY, 1964.

ies, 1888, of a copy Johnson's printed card announcing the books that he had chosen as texts for his course on logic which included Lotze's Logic.[5]

Lotze anticipated the distinction between two types of conditionals which he called *general* and *universal* that we find reflected in Johnson's Logic. It is to Johnson's credit that he thought that laws should be represented as conditionals of the second kind (Lotze's universal conditionals), while Lotze thought that there was a connection between his universal conditionals and their non accidentality.

We turn then to a more detailed consideration of Lotze's observations, and to a way of sharpening his insight into two principles that connect up lawfulness and accidentality, with the help of a condition which we shall call the *Lotze Uniformity Condition* (LUC).

Here is the seminal passage from Lotze's Outlines of Logic and of Encyclopedia of Philosophy, which raises the consideration of "unfortunate accidents".

> §32. This thought gains expression in the form of the general judgment. Such form is to be distinguished from that of the universal judgment. The latter of the form
> All S are P
> only asserts that, in fact, all instances of S have P, — for example, 'All men are mortal', — but does not tell why. Perhaps it may be on account of a combination of unfortunate accidents which have no real connection with each other.
>
> The general judgment substitutes the general concept alone for the subject: 'Man is mortal'; or it indicates by the other form, 'Every man is mortal,' that the predicate is to be considered valid, not merely of all actual but also of all thinkable examples of S; and therefore is so by virtue of this same general concept S, and not on other accidental grounds.
>
> More accurately considered, the general judgment must besides be included in the hypothetical form. For it is not the general concept S (the universal man) which is to be considered as P (mortal); but every individual, *because* he is a man. Therefore, the general form, strictly speaking, is; 'if any A whatever is an example of the universal S, then such A is necessarily P.[6]

A few remarks may make his point stand out more clearly. Lotze distinguished between *general* judgments and *universal* ones. The universal ones seem to be just the sort of generalized conditionals that can be represented by a standard universal quantifier: $\forall x(Sx \rightarrow Px)$. The second kind, the general conditional is very different from the factual one. There are two differences. The first concerns the scope of the universal quantifier. Lotze thought that

---

[5]My thanks to Elizabeth Leedham-Green, lately Cambridge University Deputy Archivist for finding the brochure, with its clear link to Lotze.

[6]*Outlines of Logic and of Encyclopedia of Philosophy*, ed. and trans. G. T. Ladd, Boston, MA: Ginn & Co., 1887.

in the universal judgment the quantifier concerned all those things which are *actual* Ss, while in the general judgment the quantifier concerns not only all actual things which are Ss, but all *thinkable* things which are Ss.

When F. Ramsey spotted this distinction as it was expressed in Johnson's logic he saw the ambiguity involved. Is the difference one of two kinds of quantifier, or one of two kinds of conditional. It looks as if Lotze and Johnson both framed the difference in terms of different universal quantifiers. And that prompted Ramsey's remark that Johnson lacked an understanding of the universal quantifier. Johnson, he said, didn't understand that "everything" means everything. However, the difference between the two kinds of judgment gets to the heart of Lotze's concept of accidental judgments. The general judgments are not accidental.

Lotze says that the predicate (mortal) holds of all and possible things in a non-accidental way that hints of explanation: the judgment he said, requires every instance of a man be mortal *because* of the uniform reason of being a man. His insight was that it was not enough for a law to be a true universal generalization: "All Fs are Gs." Something in addition to truth is needed, since it might just be an accident that the generalization is true. For suppose that indeed, everything that is an F is a G. That could have come about this way: each particular F is a G, but the reason that each F is G might be different for different Fs. If a generalization is accidentally true, then what happens is that the generalization is true, but some of its instances hold for different reasons.

What this suggests is that the truth of the instances of a law are all insured by some uniform factor. This requirement can be understood in at least three ways, explanatory, causal, or rational, — that is the provision of one explanation, one type of cause, or one reason for all of the instances. Here we shall consider only the explanatory option.

A few terms will be needed to express what I shall call Lotze's Uniformity Condition (LUC). Let us use the expression "Exp[A; B]" to say that there is some explanation of B, that is provided by A. This is an existential claim, not a reference to any particular explanation that makes "Exp[A; B]" true.[7] The particular explanation might be provided by an argument, a cause, or just a true statement. For our purposes, it doesn't matter what particular model or account of explanation one might advocate. We can now formulate the Lotze Uniformity Condition (LUC) this way: If S is the generalization (say) All Fs are G, then

$$LUC(S) \Leftrightarrow (\forall x) Exp[R; Fx \to Gx], \text{ for some } R.^8$$

Thus the universal conditional S satisfies the Lotze Uniformity Condition if

---

[7] Although we do not assume that all explanations are deductive, our use of the relation "Exp" is similar to the use in logic of the relation of deducibility "⊢", in that an existential statement is intended. "$A, B \vdash C$" means that there is a deduction of C from premises A, and B. It is existential and does not refer to any particular deduction that does the job.

[8] Here, and in what follows, we use $\Rightarrow$ to indicate logical implication, $\Leftrightarrow$ for logical equivalence, and $\to$ for the material conditional.

and only if there is some R that provides an explanation of all the instances of S.[9]

With this in place we can now construct a mini-theory of two conditions that will connect up the notion of law, accidental generalization and the Lotze condition. We say this to emphasize the fact that at this stage, we are considering something short of definitions. For any generalization S (eg. All Fs are G), and using "£" to indicate the predicate "it is a law that ...", let the conditions (1) and (2) constitute the mini theory of Laws and Accidental generalizations (LAG):

1. $£(S) \Rightarrow LUC(S)$ (for some $R, (\forall x) Exp[R; Fx \to Gx]$), and

2. $Acc(S) \Rightarrow \neg LUC(S)$ (for no $R, (\forall x) Exp[R; Fx \to Gx]$)

There are a few observations and consequences worth noting. The first condition requires that Lotze uniformity is a necessary condition for being a law: If S is a universal conditional, then some R explains every instance of it. The second condition is a necessary condition for a universal conditional to be an accidental generalization: "S is an accidental universal conditional" implies that there is no R which explains every instance of it.

Although the Lotze conditions were originally proposed for generalizations of the type All Fs are G, the uniformity condition can easily be extended to cover all generalizations of the form $(\forall x)\Omega(x)$, just as long as it makes sense to speak of the instances of Ω. Consequently the laws and accidental generalizations covered by (1) and (2) can be extended to a much broader class than just universally quantified conditionals.

2 Observations and Consequences for (LAG)

We turn now to a discussion of some of the more elementary consequences of (LAG) including those which depend on features of the concept of explanation that is used in the expression of the Lotze uniformity condition.

(i) Laws and accidental generalizations. There is an immediate consequence of (LAG) that most writers regard as central to any account of scientific laws: *No generalization S can be both a law and an accidental generalization*. That is,

For every generalization $S, £(S) \to \neg Acc(S)$.

It is a little surprising that this result can be obtained without any deeper analysis of either the notion of scientific law or explanation. If we are correct, then this account can explain why no generality can be both a law and accidental. Moreover, this account can certainly be a part of any Humean view of

---

[9]We have used the conditional $[Fx \to Gx]$ to indicate the form of the instances of a law that is conditional in form. This differs from the early discussions of confirmation in which instances were assumed to have the form $Fx\&Gx$. That led to the unacceptable consequence that logically equivalent formulations of a law would not have the same instances. Here I follow Hempel's later use of "instance" which blocks that consequence (Aspects of Scientific Explanation, the Free Press, 1965, p.341, footnote 7). We shall soon develop the uniformity condition in a way that does not require that laws have to be conditionals.

laws. There is a caveat, if Humean accounts of laws are supposed to eschew any dependence on modal notions. In that case there might be an objection on two grounds: that explanation is a modal concept (I believe that it is) and that the statement that S is a law ($£(S)$), is itself a modal statement. These observations even if correct, do not impugn a Humean account of laws. One can surely write about explanations and laws without impugning anyone's Humean credentials. Otherwise a Humean account, even if one wanted to give one, would be impossible. We think that some modals are part of any Humean's legitimate philosophical lexicon.

(ii) Laws and their instances. Condition (1) should not be confused with a requirement that is sometimes defended in the literature — that it is only laws that are confirmed by their instances. That is a view usually associated with Nelson Goodman. When coupled with the idea that the explanation relation is just the converse of the confirmation relation, one could conclude that laws explain their instances. This is a result that would make (1) obvious. The reason is just that if S is a law then there is always some R which explains all of its instances — namely S itself. This claim, even if true, is irrelevant to Condition (1), because the notion of "instance" that Goodman used is different from the notion of "instance" that is used in (LAG), as was already noted.[10] Consequently, Condition (1) gets no support from the old claim that laws explain their instances. What (1) says is that if S is a law, then there is *some* R that explains all its instances (in our sense of "instance"). It does not require that the uniform explanation be given by the law S.

(iii) The converses of the conditions of (LAG). We have not assumed the converses of (1) and (2), though both raise interesting possibilities worth exploring. According to the converse of (1), if there is a uniform explanation of each of the instances of a regularity, then that implies that the regularity is a law. The result would be that for any generalization S, S's being a law ($£(S)$) and S's satisfying the Lotze condition (LUC(S)) would mutually imply each other. This looks initially attractive since it provides something like a definition of the predicate "is a law". However we shall note below in our discussion of Reichenbach's example of golden cubes, that in combination with some assumptions about explanation, it has some moot consequences. We shall take this up when we turn to the consequences of (1) and (2) when they are combined with some additional assumptions about explanation.

Thus far we have considered consequences of the theory (LAG) that do not rely upon any properties of the notion of explanation used in the expression of the Lotze Uniformity Condition. We turn next to a number of consequences that are obtained with the help of some general assumptions about explanation.

(iv) All laws are true. That is, $L(S) \Rightarrow S$. This follows from (LAG) together with the assumption that explanations are factive. By "factivity" we mean that for any A and B,

$$Exp[A; B] \Rightarrow A, \text{ and } Exp[A; B] \Rightarrow A.$$

---

[10]Footnote 10

That means that ,"There is an explanation that A provides for B" implies A as well as B. This factivity condition is two-sided. The usual examples, such as "Richard knows that P" implies P are so-called one sided factives. In the case of explanation, both propositions are implied. In short, there is no explanation of B that is provided by A, unless both A and B are true. Here is the straightforward proof that "It is a law that ..." is factive.

Let S be the generalization $(\forall x)\Omega x$. Suppose now that $\mathcal{L}(S)$. By (1), we have LUC(S). Therefore $(\forall x)Exp[R;\Omega x]$, for some R. By factivity, $Exp[R;\Omega x] \Rightarrow \Omega x$. Therefore $(\forall x)Exp[R;\Omega x] \Rightarrow (\forall x)\Omega x$, for some R. The antecedent is just LUC(S), and the consequent is just S. Therefore $L(S) \Rightarrow S$. Thus the result is this: the sentential operator "It is a law that ..." is factive.

(v) Explained generalizations. The following result answers a simple question: What happens if a generalization is explained? The answer is surprising, but, I think, welcome. If "Exp" is closed under implication, then if there is some explanation of a generalization, then that generalization satisfies the Lotze uniformity condition, and consequently it is not an accidental generalization. The assumption of closure for explanation is admittedly controversial, though I have defended that requirement elsewhere.[11] We shall assume, in this case, that Exp is *closed under implication*. That is, for any R, A, and B,

$$\text{If } A \Rightarrow B, \text{ then } Exp[R;A] \Rightarrow Exp[R;B].$$

Briefly, if A implies B, then "There is an explanation that R provides for A" implies that there is an explanation that R provides for B.[12] The reason is fairly direct. Suppose that S is a generalization, say $(\forall x)\Omega(x)$. Since $(\forall x)\Omega(x) \Rightarrow \Omega(x)$ for all x, we have, by closure of Exp, that $Exp[R;(\forall x)\Omega(x)] \Rightarrow Exp[R;\Omega(x)]$ for all x. That is, $Exp[R;S] \Rightarrow LUC(S)$. However by (2) of (LAG), LUC(S) implies $\neg Acc(S)$. Consequently, $Exp[R;S] \Rightarrow \neg Acc(S)$. That is, if any generalization has an explanation, then it is not accidental. Lest the assumption of the closure of explanation be thought too controversial, we note that the same conclusion can be obtained with the use of the much weaker assumption of *restrictive closure*. We shall say that an explanation relation satisfies restrictive closure if and only if:

(RC) If S is a contingent generalization and $S*$ is any instance of S, then for any R, $Exp[R;S] \Rightarrow Exp[R;S*]$.

The argument now runs this way: Let $(\forall x)\Omega(x)$ be a contingent generalization and suppose that $Exp[R;(\forall x)\Omega(x)]$. Let $\Omega*$ be any instance of $(\forall x)\Omega(x)$. By restrictive closure, $Exp[R;\Omega*]$. So there is a uniform explanation (namely R) of all the instances of the generalization. Therefore $(\forall x)\Omega(x)$ satisfies the condition (LUC). Consequently, $(\forall x)\Omega(x)$ is not accidental.

---

[11] "Explanation and Modality", Typescript.

[12] This form of closure is weaker than it may appear. It does not require that there is an explanation of A which is also an explanation of B. The explanations may be different. Indeed the deducibility relation is a case where for different A, and B, the closure condition holds. Nevertheless any specific proof (deduction) of A from some C will not be a proof (deduction) of B from C.

There is an interesting moral that can be drawn from this result. It is a widespread though not universal belief that explanations are important and are sought, and prized. Why is this so, assuming that it is? One answer provided by the preceding result is this: It is one thing to have a generalization that is true. Roughly speaking, that tells us the way the world works. If the generalization is also a law, then by (LAG), that tells us even more about the way the world works. However, even if a true generalization is not a law, nevertheless, an explanation of it tells us that the generalization satisfies the Lotze uniformity condition, and that consequently, it is not accidental. That tells us more than that it is just a true generalization. There are accounts of laws that we shall mention below, that yield a stronger conclusion that the present one. They claim that explanations of generalizations imply that those generalizations are laws.

3 Laws as bound with explanation

According to an older, but mainstream tradition, explanations (whether deductive or probabilistic) are dependent upon some account of law. Accordingly, any explanation must involve an essential use of one or more laws; probabilistic laws for probabilistic explanations, and nonprobabilistic laws when the explanation is non-probabilistic. The kind of explanation will determine the kind of law that is needed. The traditional program was that accounts of explanation require that some account of laws should already be in place. Our mini theory (LAG) involves a serious departure from that program. It requires that some concept of explanation be already in place in any account of laws. The notions of lawful and accidental generalizations, on our account of them, rely on explanation. I think that the proper conclusion is not that one of the two concepts is in some sense "prior" to the other. The proper conclusion is that the concepts of laws and accidental generalizations on the one hand, and explanation on the other, are interwoven, and have to be developed in tandem.

In this connection there are two accounts of laws and accidental generalizations that deserve closer study because they also obtain interesting results by exploiting a connection between laws and explanation. One is a recent account by John Carroll, and the other a less recent account by Richard B. Braithwaite. In "Nailed to Hume's Cross?" Carroll offers an account of laws and coincidences, according to which,

> "P is a coincidence if and only if there is no Q such that P because Q" (73), and "P is a law of nature if and only if P is a regularity caused by nature." (74).

He explains that by "caused by nature" he means "that the law is true *because* of nature" (emphasis by J. Carroll); not that nature is somehow a causal agent. The intention is that "cause" should be understood in the explanatory sense that "cause" can have. This is a very interesting incorporation of explanation into an account of laws. The key notion of a regularity that is caused by nature is supposed to convey that the regularity is true because of *nature itself*. As Carroll says,

"Lawhood requires that nature itself — understood as distinct from anything in nature or the absence of something from nature — make the regularity true."(74)

This would seem to imply that all laws are explained regularities — a very significant claim. Moreover all the laws have at least this one explanation — nature. Those of us who look to theories for explanations of laws, and who also think that there may be different explanations (when possible) for different laws, may be disappointed because that kind of specificity isn't mentioned. It is easily supplied, and ought to be regarded as the fuller account Carroll intended.[13]

There are some very nice features that this theory yields. For example, it yields the nice result that laws cannot be accidental regularities, and it makes it clear that it is explanation that makes the difference between those regularities that are laws, and those that aren't. Clearly this is a theory that deserves further development and refinement.

Richard Braithwaite has also offered an account of laws and accidental generalizations which exploits their relation to explanation.[14] He used the notion of well-established *deductive systems* which are algebraic structures associated with theories, that are neither axioms for the theory nor are they Tarskian theories (sets closed under implication). and he defined laws of such systems in such a way that if h is an hypothesis in a deductive system $\mathcal{D}$, and h is explained in a deductive system $\mathcal{D}*$ which properly includes $\mathcal{D}$, then h is a law of the system $\mathcal{D}*$.[15] That is a seemingly stronger result than the one that we obtained: that any explained generalization is non-accidental. However, the results are nearly the same, given that Braithwaite regarded accidental generalizations as those for which there is no established scientific deductive system in which the generalization appears as a consequence (304). Thus he could have equally well concluded that under his conditions, the generalization is non-accidental. There are a few significant differences that ought to be noted: (1) For Braithwaite, laws are always part of some established deductive system $\mathcal{D}$ that is associated with some theory, whereas the account in (LAG) seems to consider laws independently of some theory or deductive system. Another way of indicating the difference is that for most accounts of scientific laws, the question of whether P is a law of $\mathcal{D}$ is determined by two clauses: (1) is P a law, and (2) is P a member of $\mathcal{D}$. Braithwaite's idea is that whether P is a law or not is concerned only with whether (2) is true for some deductive system $\mathcal{D}$. In focusing on the problem of whether a law is a member of some deductive system, Braithwaite may be right.[16]

---

[13]This is clear from private correspondence with Carroll.

[14]*Scientific Explanation*, Cambridge University Press, 1953.

[15]An account of Braithwaite's notion of a deductive system, laws of deductive systems, and explanations of laws in a deductive system involves several conditions o the definite inclusion of one deductive system within another. A recent discussion can be found in A.Koslow, "The explanation of laws: some unfinished business." in the *Journal of Philosophy*, 2012.

[16]A good example of this kind of account of laws that stresses the deductive role that

That insight however can be incorporated into (LAG) by adjusting the Lotze uniformity condition. Instead of the requirement that some R explains all the instances of a generalization S, we require the more refined condition that there is some theory T that explains all the instances of a generalization S. Call this the theoretical version the Lotze Uniform Theory condition: (LUTC). That is, for any generalization S,

$$LUCT(S) \Leftrightarrow (\forall x) Exp[T; Fx \to Gx], \text{ for } some \text{ theory } T.$$

This seems to us to be a natural refinement of the version that only asks for some explanation. After all we are presently interested in the explanations of contingent generalizations and their instances, and that is rightly thought to be the work of theories. The idea then is to use the theoretical version of the uniformity condition in the expression of our mini-theory (LAG), replacing "(LUC(S)" by "(LUTC)" so that our mini-theory is now given by (LAGT):

(1*) $L(S) \Rightarrow LUTC(S)$ (for *some* theory T, $(\forall x) Exp[T; Fx \to Gx]$), and

(2*) $Acc(S) \Rightarrow \neg LUTC(S)$ (for no theory T, $(\forall x) Exp[T; Fx \to Gx]$.

We shall return to a discussion of this version of the uniformity condition when considering the question of how well this account of the distinction between laws and accidental generalizations squares with some familiar examples in the literature.

4   Gold spheres, Rusty Screws, and Rocks with Iron

We will consider those examples, first with the use of the uniformity condition that requires that there be some R that explains each of the instances of a law (LAG), and then the version of the uniformity condition that requires that there is a theory T that explains all the instances of the generalization (LUTC).

There are three classic examples of regularities that fail to be laws. There is H. Reichenbach's "All gold cubes are less than one cubic mile in volume." (G), E. Nagel's "All the screws in Smith's care are rusty." (N),[17] and C. Hempel's "All the rocks in this box contain iron." (H).[18] They all seem to reach the conclusion that none of the examples are laws by appealing to the unexplained claim that they are accidental generalities, and consequently not laws. We shall show that the same conclusion about their failure to be laws is also available on our account.

The most interesting example of the three is H. Reichenbach's example of two statements of the same logical form:

(U) All uranium cubes are less than one cubic mile in volume.

(G) All gold cubes are less than one cubic mile in volume.

---

a law plays within a scientific theory, but is otherwise very different from Braithwaite's is the richly detailed recent one of John T. Roberts in *The Law-Governed Universe*, Oxford University Press, 2008.

[17] *The Structure of Science*, Harcourt, Brace & World, Inc., 1961, pp. 58-59.

[18] *Philosophy of Natural Science*, Prentice-Hall, Inc., 1966, p.56.

The standard assessment is that (U) is a law and that (G) is not.

One reason given for (U) being a law is that it follows from the laws of Quantum Mechanics, while (G) is usually considered to be obviously accidental.[19] The mini-theory (LAG) doesn't yield the conclusion that (U) is a law (It does not provide a sufficient condition for being a law), nevertheless it does yield the important result that (U) is not accidental. The reason is that there is an R, namely Quantum Theory, that explains all the instances of (U), and that is enough, as we have seen, to guarantee that it is not an accidental regularity.

The verdict for (G), that is based on (LAG), is the familiar one: (G) is not a law. This is not so for a version of the uniformity condition that is based on there being a reason R, that is not necessarily an explanation, for all the instances.[20]

To see that (G) isn't a law, recall that according to (LAG), we have $L(G) \Rightarrow LUC(G)$. Moreover, according to LUC(G) there is a uniform explanation why each lump of gold is less than a cubic mile in volume. But there isn't such a uniform explanation for gold (as there was in the case of uranium where the uniform explanation was provided by Quantum Theory). So, LUC(G) is false, and consequently, (G) is not a law. The same conclusion follows, using the theoretical refinement of (LUC), since there is no theory that explains why the volume of each lump of gold is less than one cubic mile.

The familiar examples that Nagel and Hempel provided as non-laws, (N), and (H), also turn out to be non-laws on our account. The argument is the same as for (G). Our account requires that $L(N) \Rightarrow LUC(N)$. Now LUC(N) is false since there isn't any uniform explanation of why each of the screws in Smith's car are empty. Now you might plausibly think that there could be a uniform reason why all the screws were rusty — say perhaps that some fiendish mechanic replaced all the screws by rusty ones. That would be a reason why all of them were rusty, but it would not be an explanation, and it is an explanation that is required. The same verdict would be given on the basis of the theoretical version of uniformity— that there was some theory that explained why each screw was rust. But there is no theory about the screws in Smith's car. Similar verdicts hold for Hempel's "All the rocks in this box contain iron." There's no theory for those rocks.

In laying out the claims of (LAG) and (LAGT), we do not by any means

---

[19]Though it is possible to accept these judgments, the argument is sometimes given that (U) is a law because it follows from laws. That however is not a good argument. It is not generally true that every logical consequence of laws is also a law. As for (G), it is usually thought to fail to be a law because it is assumed that it is accidental.

[20]If one based the uniformity principle (LUC) on a uniformity condition such that if a contingent generalization is accidental, then there is one reason, which needn't be an explanation) for all its instances, then there are scenarios in which the gold example (G) would not be accidental. Here's one such scenario: The total amount of gold in volume is less than one cubic volume. That singular fact would imply all the instances of (G), but wouldn't explain them. Consequently (LUC) would be false, and consequently, (G) would not be accidental. As we have indicated, our theory does not use such a weak version of the uniformity condition, but requires that there be a uniform *explanation* of all the instances of the generalization.

think that this is all that can or should be told about laws and accidental generalizations. We think it is just part of a fuller story. As it stands so far, it does not cover laws that are schematic, laws that are not obvious regularities, and so forth. It remains to be seen whether the present story can be extended to cover those cases. It is true that on the basis of this account so far, if a generalization is explained, then it is not accidental. The obvious corollary is that there is no explaining an accidental generality. That however may not diminish their scientific importance, or utility, or even their use in explaining other regularities.

Acknowledgements

It is a great pleasure to write in honor of Pat Suppes. I first came to know his work when I was a graduate student at Columbia, and came across a copy of his dissertation *The problem of action at a distance* (supervised by E. Nagel) on deposit in the philosophy library. It was a revelation. The combination of historical accuracy, formal precision, elegance and intrinsic interest had a profound influence on me. When I had a year fellowship to study philosophy anywhere I wanted, I made a bee-line directly to Stanford, then Harvard, and Cambridge. My early work on measurement, and later work in logic bear his imprint, if not his imprimatur. Thanks Pat for showing me the way.

I am grateful to Alberto Cordero for his enthusiasm and interest about the central idea of the paper early on, to Hugh Mellor, and to members of my Graduate Center CUNY seminar on scientific laws, and to two anonymous referees for helpful comments.

BIBLIOGRAPHY

[1] Braithwaite, R.B.B, *Scientific Explanation*. Cambridge University Press, 1953.
[2] Carroll, J., Nailed to Hume's Cross. In: *Contemporary Debates in Metaphysics*, Eds. J. Hawthorne and D. Zimmerman, Blackwell, 2008.
[3] Hempel, P. *Aspects of Scientific Explanation*, The Free Press, 1965.
[4] Hempel, P., *Philosophy of Natural Science*. Prentice-Hall:NJ, 1966.
[5] Johnson, W.E., *Logic Part I*. Cambridge University Press, 1921, Reprinted by Dover Publications, NY, 1964.
[6] Johnson, W.E., *Logic, Part III, The Logical Foundations of Science*. Reprinted by Dover Publications, NY, 1964.
[7] Koslow, A., The explanation of laws: some unfinished business. *Journal of Philosophy, Aspects of Explanation, Theory, and Uncertainty: Essays in Honor Of Ernest Nagel*. Eds. B.Berofsky and I. Levi, **8/9**, August/September 2012.
[8] Koslow, A. Explanation and Modality. Typescript.
[9] Lotze, H., *Outlines of Logic and of Encyclopedia of Philosophy*, ed. and transl. G.T. Ladd, MA: Ginn & Co., 1887.
[10] Nagel, E., *The Structure of Science*. Harcourt, Brace & World, Inc., 1961.
[11] Roberts, J.T., *The Law-Governed Universe*, Oxford University Press, 2008.

Modeling Causality
JEAN-YVES BEZIAU

ABSTRACT. This paper is a methodological exercise, applying model theory to causality. We start by explaining that model theory does not reduce to a formal tool, that it is an interesting and deep philosophical approach. We then develop a framework where causality is a binary relation between objects considered as events. From this perspective we examine the so-called principle of causality and we also discuss other possible axioms for the relation of causality analyzing their significance and import. We end by a case study: citation in research papers viewed as a cause-effect phenomenon

Dedicated to Patrick Suppes (1922-2014)
The Last Cowboy of Thought

$a \hookrightarrow b$

La cause la plus profonde se son malheur c'est qu'il pensait
qu'il n'y avait pas de fumée sans feu,
et quand la poudrière de sa voisine n'était pas allumée,
il avait peur que que tout s'envole en fumée.
Baron de Chambourcy

0 Toward a Theory of Causality

"Causality" is a word and there is an idea (notion, concept) corresponding to it. Furthermore this idea may apply to reality, it can describe and explain something in the world: the physical world, the biological world, the economical world, the psychological world, etc. This trinity word-idea-reality is not proper to causality. To the word "mud" corresponds a reality in which we can flounder, that in turn can be characterized by an idea, *earth mixed with water* (cf. Plato's *Theaetetus*). But we don't necessarily have a trilogy for any substantive, because the distinction between idea and reality is not always clear, think for example of infinity.

In this paper we intend to study the articulation of such kind of trinity using model theory taking as an example causality. This can be seen as a methodological exercise. But this is not just a game. We intend to reach understanding, both of causality and the methodology of model theory. We are using the expression "methodological exercise" to stress that our interest is not only for causality but also for the methodology, be it apply to causality or other notions.

One may claim that causality has no proper meaning. Such an argument may apply to Guilt, the G-spot or God. But who shall we blame for the lack of meaning? The word or the idea? It is difficult to blame a word. Words are used to express ideas. It is true that many words reflect the confusion of our thought. If we want to inquire what causality is, we certainly cannot stick to the word, developing a purely descriptive approach, the way the word (and its linguistic variations, "causalité", "causalidad", "causalità", "Kausalität", etc.) has been used from Hume to Judea Pearl [13].[1] The notion of causality, as many notions, is relatively independent of the word. Aristotle, famous for his theory of four causes, was using the Greek word "aitos" which has no linguistic relation with the neo-Latin word "cause". Despite this change of language and even if we do not agree with Aristotle's theory, there is no doubt that the target of these two words is the same: the cause of something is how and why this thing happens. In other cultures this notion has also been thought with different words. In Chinese for example, the word is "gù':[2]

Taking in account the limited import of the word, one may want to develop a descriptive approach of the idea behind the word, ranging from Aristotle to

---

[1] In English "causality" and "causation" are both used. In some sense "causation" is better because, according to the Cambridge dictionary, it means "the process of causing something to happen or exist" by contrast with causality defined as "the principle that there is a cause for everything that happens", involving a confusion between causality and the principle of causality. The reason why however we are using "causality" is due to its variations in other European languages where there are no equivalent to "causation".

[2] Thanks to Zhenzhen Guo for this indication and references to the *Mozi*.

David Bohm [7], but this is also limited. What we can do is to develop a *theory* of causality, trying to see how we can think about causality in a systematic, clear, meaningful and useful way. Such a theory cannot be absurdly normative; in this case there would be no reason to still use the word "causality", it would be a theory of something else. We have to find a good balance between descriptivity and normativity. Tarski has developed his theory of truth in this spirit [18]. To do so we have to take in account the meanings which have been attached to the word "causality" and the notion of causality.

However we have to be aware that in exact sciences sometimes the theory of a notion is quite remote from its usual meaning. Consider for example the case of time. This is maybe because the very nature of time is quite different from the vulgar idea of time. The way to an advanced scientific theory of a notion is not necessarily a straightforward road leading us deeper and deeper, higher and higher. It can be a tortuous path driving us to an unexpected notion we were not able to dream of. But in this case the reason why we use the same word for the remote reached point and the starting point is because the starting point was the starting point.[3]

So perhaps by developing a theory of causality we may reach something quite different from the common notion, because the very nature of causality is remote from the basic idea of causality. Someone may not believe in the *very nature* of causality, or *very nature* of time. This can indeed be just a way of speaking. We can instead speak of a *more sophisticated* notion, *more useful* or/and *applying better to reality*.

There is a difference between on the one hand a notion like reasoning, and on the other hand a notion like star. In the case of star, there are stars in reality and we want to develop a theory of this notion that describes, fits, grasps, explains this reality. We can say the same about a phenomenon like fear. In the case of reasoning, we may want to develop a normative theory. Normative in the sense not only that it is a better way to think about reasoning different from the confused ordinary idea, but normative in the sense that this is how reasoning should be performed. There is in this case an interaction between the theory and reality. The theory is transforming reality. Reasoning is not an independent reality.

A notion like time is more like the notion of star or fear, although it is more complex. In the case of causality it is more ambiguous, because one may argue that causality is just a way to think about reality. However "a way to think" of reality can be considered as a reality not in a noumenal sense, but in a phenomenal sense, to use Kant's terminology. Saying it is a reality basically means here that it has a nature which is independent of what we can think about it. A philosopher like Schopenhauer has presented a phenomenal theory of causality in a neo-Kantian sense: causality for him is one of the four roots of the principle of sufficient reason [16], this principle being a phenomenal reality that we cannot transform, but that we can describe more or less adequately.

---

[3]We have however to be cautious not to commit a Columbus confusion: natives of America have been called "Indians" but they have nothing to do with India. Having in view an idea we may reach another one ...

One may argue that causality does not properly exist, that it is a primitive and pre-historical notion that has no sense nowadays, that the cartography of our world of ideas has so much changed that it is as absurd to talk of causality in the contemporary world as to talk of Prussia or Aether.[4] But one may also sustain that causality is a notion that can be transformed, that we can develop a normative theory of causality that makes sense. Trying to develop a theory of causality is a way to see if past and future theories of causality are possible or not, if this notion has ever made sense or will ever make sense. Recently this notion has been discussed quite a lot, mainly by philosophers (see e.g. [1]), perhaps because it is a general notion that at this stage only philosophers can think about and hopefully save. Thinking about causality is a challenge for philosophy, but we have to see if it is possible to develop intelligent philosophy going beyond wordy discussions and disputes.

Our work here about causality is an exercise in philosophy. We will see how it is possible or not to develop a theory of causality using model theory.[5] We don't think that philosophy has necessarily to be developed using model theory, or that philosophy consists of developing theories. But what we want to show is that by so doing we may clarify the notion of causality.

1 The general framework: causality-effects as a binary structure

1.1 The methodology of model theory and causality

Let see how we can think about causality. Our idea here is to use model theory in a simple but authentic way. Model theory is a meta-theory in the sense that it is a theory to develop theories. Model theory can be applied to mathematics: theories of numbers, of spaces, of groups, fields but also to any non-mathematical field ... It is true that model theory has been mainly developed for mathematical theories and for this reason is strongly associated with mathematics. But the originator of model theory, Alfred Tarski,[6] had a

---

[4]One century ago Bertrand Russell wrote: "In the following paper I wish, first, to maintain that the word "cause" is so inextricably bound up with misleading associations as to make its complete extrusion from the philosophical vocabulary desirable." [15]

[5]Our paper is written in such a way that it can be understood by someone who knows nearly nothing about model theory. But at the same time it can be of interest for model-theorists. We don't present here a simplification of model theory that could be boring for the specialist, but rather a view of it, not too technical, stressing its philosophical value. Let us also emphasize that it is possible to find in the literature some approaches of causation quite close to the framework we are presenting, events related by a binary relation of causality, see e.g. [11]. Maybe in these approaches the authors are using "informally" model theory, but our aim here is to explicitly use model theory.

[6]Someone may argue that model theory was existing before Tarski invented the expression "model theory". But here there is a strong relation between the expression and the topic. Maybe it would be better to say that before Tarski coined the expression and develop the corresponding theory, this was just the pre-history of model theory. Hodges, in his seminal paper "Truth in a structure" [10], tries to explain why it took so-long to give the now standard first-order definition of truth.

general perspective and had interest to develop this theory for sciences like physics and biology. With Patrick Suppes and Leon Henkin he organized a conference on *Axiomatic Method* in December 1957 at Berkeley. This was a preliminary step for the launching of the series of congress *Logic, Methodology and Philosophy of Science* (LMPS), the first LMPS being organized by Suppes at Stanford University in 1960. Today model theory is not an important part of LMPS and of philosophy of science in general. What has become increasingly popular is "modelization", the exact meaning of this word not being very clear (see [12]). We want to show here how model theory can fruitfully be used, and can be seen as a good way to modelize.

The other reason why model theory is associated with mathematics is that it is rather a mathematical theory, although this point is not completely clear: it appears more mathematical for someone outside of mathematics than for a mathematician. The truth is that it is a very abstract theory that would be better classified as philosophical or metaphysical.[7] The basis of model theory is a structural point of view according to which we have objects and relations between objects. Some objects given together with some relations form a *structure*.[8] Here objects are conceived in a very general sense. They are not necessarily objects like a stone or a number, it can be an emotion, rain, any phenomenon. What is important in model theory is the two level perspective: objects on the one hand, relations on the other hand, objects being understood through the properties of the relations between objects.[9] These properties are characterized by some axioms.

A group of axioms is called *a theory*.[10] A theory has different *models*, structures obeying the axioms, structures which can be more or less different.[11] Sometimes one may want to reach unicity, i.e. all models are the same, in this case the theory is said to be *categorical*. This option is important in particular when one has in view the objects of the structure, for example if one wants to characterize natural numbers. If a theory of natural numbers has very different models this means that we are not succeeding to characterize what is a natural number. This is in fact exactly what happens: in the theory called "Peano arithmetics" we have non-standard models with strange objects. This is an important result due to Tarski himself.

---

[7]The same can be said about first-order logic, model theory being one fundamental aspect of first-order logic.

[8]The notion of structure can be considered as a primitive notion and a set can be considered as a limit case of structure.

[9]These objects can be anything; they can also therefore be relations. But at this stage they are considered as objects by contrast to the relations which are considered to study them in the structure. In model theory there can also be functions, but we will not discuss this notion here. Functions can be considered as a particular type of relations.

[10]Here "theory" has not the same meaning as in "model theory", or in "number theory", a detailed analysis of the words "theory" and "model" can be found in [5].

[11]In model theory, we have on the one hand "theories", on the other hand "structures". These structures can in general be called "models", considering that a structure can always be a model of some theory.

If one has in view relations rather than objects, it is not necessarily important to look for categoricity. For example the notion of order is characterized by the axioms of antisymmetry and transitivity and it has many different models describing all the variations of this notion that we don't want to eliminate. This is not incompatible with the fact that one may want to focus on a particular notion of order and to categorically catch it, for example the notion of dense order. The notion of density can be easily captured because it is easy to find an axiom for density; nevertheless this axiom is not enough to get categoricity. To have a categorical theory of dense order we must choose between secondary features, in particular having or not a first or a last element. For this reason there are different categorical theories of dense order, the most famous being the one having for models exactly the rational numbers. These theories are incompatible in the sense that they don't have common models.

Developing model theory for causality we have to be aware of this distinction. The situation of causality can perhaps be compared to dense order in the sense that we may want to categorically characterize causality, to catch the very idea of causality. But the model theoretical perspective clearly shows us that we can have a less categorical approach, having like for order relations a central set of axioms extending in many different specific theories of causality incompatible with each other, which is not necessarily a problem. A relation of order can be discrete or dense, these are incompatible properties, nevertheless these two theories have a common ground, the two basic axioms for order.

The fact that we have different models of the same theory of causality may be seen as the variations of the notion of causality according to various "interpretations". From a structuralist point of view, if two structures are different, this means that the objects of the structures are different. Different models of the theory of causality can therefore be seen as different fields to which the notion of causality applies.

1.2 The relation of causality in a model-theoretical perspective

The simplest non trivial situation in model theory is when we have only one kind of objects and one relation between these objects, a binary relation. It is possible to show that more or less artificially any structure can be reduced to such kind of simple structure. This kind or reductionism is a technical result that can be useful for general metatheorems, but not necessarily interesting if we want to be close to the concepts we want to characterize. It happens that we can think about causality quite naturally using such a simple structure.

We consider structures with a binary relation on a set of objects, that we call *relation of causality* or *causality* for short and that we can symbolically represent by the capital letter "C". Given two objects a and b, we write aCb and this can be read as "a is in causal relation with b", or more simply "a causes b". In the second formulation we use a verb, this is not a problem at all, model theory is much more general that people usually think having in mind some specific applications of it. The notion of relation is something very abstract, and can be an (inter)action between two objects and many other "things".

When we use the letter "C", this can be viewed in two ways: an abbreviation and an abstraction. These two ways are not opposed but they also do not reduce one to the other. It is in fact not just an abbreviation, for this reason, and to avoid this ambiguity it is better to use another sign which is also more suggestive and symbolic like "$\hookrightarrow$". Then we write $a \hookrightarrow b$, expression we read as "a causes b".

This is quite similar as what is done with the relation of order, when we write $a < b$. Note that, except the rather complicated expression "a is in order relation with b" (which does mot clearly appear as different to "b is in order relation with a" - an unfortunate ambiguity especially for the case of an antisymmetric relation), there is no direct reading of this expression, because expressions like "a is inferior to b" or "a precedes b" are already interpretations of the relation of order.

Another example of relation is the relation of consequence. We have the expression $a \vdash b$ which is read as "b is a consequence of a a". The relation between the two objects is a relation of *consequence* in the same way that in the other cases we have relations of *causality* and *relation of order*. But then the word "consequence" is transposed into the object which is at the right of the relation, which is qualified as a "consequence".

In the case of causality the word naming the relation can be transposed on the left, to qualify the object at the left: "a is a cause of b". In the case of the relation of order, the word naming the relation is transposed neither on the left nor on the right. In the case of causality, we have also a name for the object which is at the right, it is called an *effect*. When $a \hookrightarrow b$ we say that b is an effect of a. In the case of a consequence relation, we can have also a word to qualify the other side of the relation, $a \vdash b$, we can say that a is a hypothesis (for b).

Like in the case of consequence there is a kind of disparity since the name of the relation is transposed only on one side. But what is important is that from our model theoretical perspective the objects of both sides of the relation of causality are of the same nature. This homogeneity corresponds to a framework not suitable for Aristotle's theory of causality or other theories. In the case of Aristotle's theory, on the one hand it would be difficult to argue that the four different causes are of the same nature and on the other hand that the effect, product of these four causes is of the same nature as these causes.

The fact that we use different names can be interpreted as meaning that these objects may have two different roles corresponding to their "positions". This way of speaking is usual in natural language for various situations: someone may be at the same time a son and a father. In our ordinary language it is not however clear that the words "cause" and "effect" just correspond to roles, maybe the idea is that they also correspond to two different kinds of objects. This is not really clear and in fact it is not clear also what kinds of objects correspond to either of these words. The word "object" would not even be used, there are no proper names for the things corresponding to these words. The way of naming here is a kind of reduction of something to its function. We will try here to construct objects behind the words, through a structural approach, which is not a pure functionalism.

1.3 Events

In model theory, given a structure, it is quite normal to speak of "objects" or "elements" of the structure. The word object is used in an abstract sense not corresponding to the common use of the word and when were are studying a particular class of models, we give to these objects the name of intended interpretations of these objects.

Here we will choose to call the objects "events". Let us keep emphasizing that since we are making a theory, the word "event" and correlated notions ("cause", "effect", "causality relation") are not considered in a purely descriptive sense. But we are trying to take a word which is as neutral as possible, and we are not dealing only with the word "event" but also with the notion associated to it.

Other words/notions are possible like "phenomenon" or "process", the idea being to choose moving things rather than static things. It makes sense to say that the work of the carpenter is the cause of the table. But since the idea of our model-theoretical framework is to have on both sides of the relation of causality the same kind of objects, we can instead say that the work of the carpenter is the cause of the production of the table, production of the table being an event, and the table a byproduct that we assimilate with this event.

Event can be seen as a quite general notion. Physical atomism reduces everything to atoms considered as small indivisible material objects. In the case of logical atomism, we have atomic propositions, corresponding to facts, which cannot be decomposed into other propositions, corresponding to more elementary facts. One may want to defend an "event atomism" considering that the world is based on some atomic events that cannot be divided. The option is not incompatible with our present theory, this is a particular case of it we are not especially defending here, but we leave this option open.

To further develop such a theory it would be useful to introduce in our structure some functions composing events into other events. According to our simple structure, the events are not necessarily considered as indecomposable, and also we do not suppose that there are such indecomposable events. We defend here a structuralist approach: what is important are the relations between events, and these relations are understood through a single notion, the notion of causality.

Let us emphasize that our perspective is not absolutist: though this structure causality-events is very general, it is only a way to look at the world. We can say that our theory causality-events is *universal*, and it is an important feature of it, in the sense that it encompasses all phenomena, but it is *relative* in the sense that it is just one possible way to look at them.

The standard meaning of "event" is: " anything that happens ". This sounds very general and it makes sense to say that anything happens, so that anything can be considered as an event. One may consider that an event is an interaction between some "things", like a stone thrown at a window. Here we have two "things": stone and window. But in our present theory we will not decompose events in more elementary things. Putting events in the first place means that we are considering what is happening and only that, in the same way that in propositional logic one does not decompose the proposition in other kinds of entities.

One may consider than the relation of causality is of a similar type as the relation of consequence rather than the relation of order, for a reason of multiplicity. Although we have talked about the consequence relation through an example such as $a \vdash b$, generally we have something like $a_1, ..., a_n \vdash b$, i.e. from various hypotheses we reach a conclusion. For causality we can also naturally consider that we have a multiplicity of causes leading to an effect: $a_1, ..., a_n \hookrightarrow b$. But we can consider that the multiple causes $a_1, ..., a_n$ together correspond to a single event. Of course it would be useful to have a function combining multiplicity of events into one single event, like we have connectives combining a multiplicity of propositions in one single proposition. But we can work first at a more abstract level without such functions. This analysis can also be applied on the right of the causality relation, we can consider, like in the case of multi-conclusion consequence relation, that there are many objects on the right, a cause having several effects. These several effects can be considered as one effect, "product" of these different effects, following the above analyzis for cause.

2 The principle of causality

2.1 One fundamental principle

Many questions that have been discussed about causality can quite easily be reformulated in our framework. We can analyze the new dimension that it shed on them but we also always have to be careful about the specificity and limitation of this perspective.

There is the question to know if everything has a cause. This can be understood in various ways. In our approach, this proposition can be stated as "Every event is caused by another event", which corresponds to the following first-order formula: $\forall x \exists y (y \hookrightarrow x)$. This can be considered as a formulation of *the principle of causality*.[12]

This principle is quite mythological, mythology promoted by a lot of ambiguities. The word "principle" itself is rather ambiguous. The word "axiom" is more neutral and relative, especially in the context of modern axiomatic and model theory.

There exists an ambiguity that persists even if we use the expression *the axiom of causality* instead of the expression *the principle of causality*: it suggests that this is the "one fundamental" thing that describes or/and defines causality. Imagine that we were using the expression "the axiom of order" to designate the axiom of antisymmetry. This is not the general way of speaking because the idea is that the notion of order is defined at least with two different axioms: antisymmetry and transitivity. Nevertheless one could argue in favor of using the expression "the axiom of order" for antisymmetry defending a position according to which, though antisymmetry is not the only axiom for order, it is the most fundamental one. Such expression does not necessarily imply pure uniqueness.

---

[12]In our framework "Every event is caused by another event" is equivalent to "Every event is an effect".

To find one axiom for causality, this is the least we can aim at. If we find several that can also be fine. But then we have to classify the different axioms: to see if there are some more fundamental than the other ones. In the case of the theory of order, we have two fundamental axioms – antisymmetry and transitivity – and then a bunch of "superficial" axioms, that extend the theory in different ways, such as the density axiom, the first element axiom, etc. To call the formula $\forall x \exists y (y \hookrightarrow x)$, "THE principle of causality", means that this is the only one fundamental axiom for the theory of causality. This is a position than can be defended but one has to be aware of its significance.

2.2 The principle of causality and the principle of sufficient reason

"Everything has a cause" can be understood in different ways, depending on what kinds of things we are talking about: the cause of an economical crisis, the cause of a war, the cause of a storm, the cause of a illness, the cause of a depression.

There is a connection between the so-called *principle of causality* and *the principle of sufficient reason* (PSR) which is expressed as "Nihil est sine ratione". Translating it into English and putting it in a positive form we have: "Everything has a reason". This can be seen as a more general principle. This perspective has been precisely defended by Arthur Schopenhauer in his PhD *The fourfold root of the principle of sufficient reason*, where causality appears as one of the fourth forms of this principle, the one having to do with physical objects. Causality is one reasons among others, related to a specific field of objects. The theory of Schopenhauer is quite interesting. Without following necessarily the same division he is operating, one may want to argue that causality applies only to a certain class of objects.

In a sense this is what we are doing, because for us causality does apply only to events, not to any kind of things. But events are not really for us a part of reality, they rather are a partial way to look at reality. So our theory is not against applying causality to mental or psychic phenomena as is Schopenhauer's theory (another form of the PSR different from the principle causality applies to mental phenomena according to the Danziger). Our approach is also not radically against applying causality to propositions / reasoning, saying that some hypotheses are causes of a theorem (for Schopenhauer this is is again another form of the PSR, different from causality).

When we are looking at axioms for causality, we don't necessarily mean that mental phenomena, reasoning, physical objects are of the same type, we mean that viewed from a certain perspective, i.e. viewed as events, they obey some common axioms. Nowadays it is quite common to speak about "mental causation". It is not clear that people are aware of the confusion that can result from this expression, assimilating mental phenomena with physical phenomena. Such an assimilation can be defended by physicalists, but we can talk about causality between mental events or between mental events and physical events without being a physicalist or a dualist (cf. the mind-body problem), with no commitment about the ontological nature of events.

2.3 The variety of models of the axiom of causality

To defend the axiom of causality understood as "Every event is caused by another event", one must be able to reject any counter-example, event without a cause, and/or be conscious of what is exactly implied or not by this statement, both from a philosophical and logical point of view,

That every event has a cause does not necessarily mean that there is a first cause, an event that is the cause of everything including itself, that can be called God, if we are kind enough to consider God as an event. This is an option among others. The principle of causality by itself is therefore neutral regarding such a God.

There are different models of the theory constituted by this sole axiom: one possible model is a model of infinite chains of events, another possible model is a model with cycles, we can also have a model which is a mix of the two. If one wants to eliminate these models, he has to put some additional axioms. And philosophically speaking one has to argue for these axioms. If one does not put these axioms, one has to be ready to explain the meaning of the variety of models of the axiom of causality. What we see is that one may have different reasons to defend the axiom of causality.

When claiming that everything has a cause, one may want to exclude multiplicity of disparate causes, even admitting that in general something may have different causes. Our framework here permits to defend this idea, because according to our formulation, every event is caused (at least) by a single event. As we have explained, this event can be viewed as a conjunction of events, but not any conjunction of events can necessarily be considered as a single event.

2.4 Rejection of the axiom of causality and counter-axiom

Let us see now if we can find some good reasons to reject the principle of causality. Rejecting this principle means that there is (at least) an event which has no cause, symbolically speaking: $\exists x \forall y (y \not\to x)$. Even if we have a good example / specimen to sustain this idea, we can wonder if it would make sense to take this formula as an axiom for a theory of causality.

This axiom, which is the negation of the principle of causality, starts with an existential quantifier and moreover includes a negation. Such configuration does not prohibit this formula to be an axiom: in famous theories we have a similar axiom. In a relation of total order $\exists x \forall y (y \not< x)$ means that there is a first element. But this would be strange to have only existential axioms, in particular for a theory describing reality, because the spirit of such "empirical" theory is to describe some general features of reality, and generality means universal quantification in the first place. Suppes and Chuaqui have shown that classical physics can be axiomatized by a theory only with universal quantifiers (and without negation), see [8].[13]

If we are not interested to have a complete and categorical theory we may

---

[13]Having in mind that causality and events are (a certain aspect of) reality, physical or mental, it makes sense to talk of the *law of causality* alternatively to the principle of causality.

choose neither $\forall x \exists y (y \hookrightarrow x)$ nor $\exists x \forall y (y \not\hookrightarrow x)$ as axioms. In this case we have models of the theory of causality in which every event has a cause, and others in which there are events without a cause. This not necessarily a problem in the sense that these two classes of models can be seen as describing two kinds of universe of events.

Let us see examine now other axioms that can be considered for the theory of causality.

3 Other axioms for causality

3.1 Everything has an effect

Everything has an effect can be formulated in our framework as "Every event causes another event", equivalent to "Every event is a cause". This appears as the reverse of the principle of causality and can be expressed by the following first-order formula: $\forall x \exists y (y \hookleftarrow x)$. Traditionally this appears as weaker than the principle of causality according to which if something happens there must be a reason, this is no purely contingent. This reverse principle does not even has a name! Like a stray dog … But everybody understands its meaning, it is not like an unknown creature that we would encounter by surprise.

One of the reasons to deny this principle would be on the basis of identifying tininess with nothingness. But one has to be careful of the butterfly effect. Another reason is to stress the differences among phenomena, for example between thought and action. To kiss someone is not the same as thinking of kissing. Thinking may lead to action or … nothing.

One can defend the reverse principle of causality in the perspective of a "relational" philosophy: not that everything is related to everything, but something is necessarily related to something else, upstream (as a cause) or downstream (as an effect). However a weak relationalism can be based on a disjunctive axiom, every event has a cause or an effect: $\forall x (\exists y (y \hookrightarrow x) \vee \exists y (y \hookleftarrow x))$.

3.2 Reflexivity

Does it make sense to say that an event is its proper cause? If one wants to rule out this possibility, nothing easier, she just has to choose the axiom: $\forall x (x \not\hookrightarrow x)$.

For a relation of order it is also possible to choose this option, in this case we speak of a *strict order*. It is also possible to choose reflexivity: everything is *superior or equal* to itself. In the case of a relation of order, it seems that one has to choose between these two options, a choice between two relations expressed by two different signs $<$ or $\leq$. The choice of the notation expresses the choice of one option. There is no notation for a relation of order with some elements which can be in relation with themselves, and others not.

In the case of a causal relation, it makes sense to have such kind of mix models. One may think that there is an event, which is cause of itself, and that this event is unique, a God in monotheism. To think that all events are causes of themselves, would be an exaggeration. But one may sustain that there is a class of events which are causes of themselves. A good way to develop this idea is to consider events which are causes of themselves but having no other causes than themselves: $\forall x ((x \hookrightarrow x) \to (\forall y (y \hookrightarrow x) \to y = x)))$ These events can be interpreted as simulating events without causes and can permit to defend the axiom of causality in an original way.

Such events are causes of themselves and they are also effects of themselves, but there are no reason to consider that they don't have other effects than themselves. An event which is cause of itself and nothing else and which is at the same time effect of itself and nothing else would be a quite strange isolated phenomenon. On the other hand we can consider that there are events which have different causes but no effect, some kind of "dead-end" events. This makes sense in the case of generation and family.

3.3 Antisymmetry

If we consider that some events may be causes of themselves, then antisymmetry has to be formulated as follows: $\forall x \forall y ((x \hookrightarrow y) \land (x \hookleftarrow y) \to x = y)$. Otherwise we can formulate it as: $\forall x \forall y ((x \hookrightarrow y) \to (x \not\hookleftarrow y))$.

Many people would argue that causality is necessarily antisymmetric, mainly because they think that time is antisymmetric and that events are dependent of time. First let us point out that it is possible to have a theory of time which is not antisymmetric, having some circles or cycles - cf. Gödel's model of Einstein's theory [9].

On the other hand consider the following cycle of events: Spring leads to summer, summer leads to fall, fall leads to winter and winter leads back to spring. One may argue that we started with spring of 2014 and end up with spring of 2015, which is not the same spring. But we can considered a season independently of a particular instantiation of it. In the same way we can say that chicken cause eggs and eggs cause chicken, this a symmetric causality.

For the seasons one may argue that even if spring leads to fall (through summer) and fall leads to spring (through winter), a season is not an effect of a previous season, but that it "follows" the previous season. Nevertheless there are indeed many good reasons to see a more intrinsic relation between seasons. We must be cautious with argument such as "This is not causality because this is not antisymmetric". This is a kind of vicious circle argument, presupposing what we want to infer. A better argument against the above seasonable example would be to defend that causality is not transitive: spring causes summer and summer causes fall, but springer does not cause fall. However this non-transitive argument does not apply to chicken and eggs: there is nothing in-between.

Another way to defend symmetry of causality, even with antisymmetry of time, are interactive events, which are simultaneous in a certain portion of time. This is illustrated by our case study (see next section): a research paper can cause another one and vice-versa, there is a symmetric causation by simultaneous quotations.

3.4 Transitivity, Chain and Connectivity

Transitivity if clearly not an axiom for causality, although there may be some classes of events to which it applies. Let us emphasize than if we claim that an event a causes an event b and that an event b causes an event c, we can claim that a causes c even if the intermediate event b necessarily has to happen in-between. It makes sense if it is a chain of homogeneous events without outsiders, like the fall of dominoes.

Now consider that with your credit card you go to an ATM machine, take some cash and with this cash money buy some bananas in a street market where cash is not accepted. In this case we can say that your credit card is not the cause of buying these bananas, because it would not have been possible to directly buy bananas with your card. You credit card could not directly causes this effect. We can say that it is an "indirect" cause. When having a causality chain, we can distinguish between direct and indirect causes. An indirect cause is a cause such that there is necessarily an event in-between. But we have to be careful to make the distinction between dominoes and bananas.

The formal definition of chain for causality is the the same as for relation of order, even if we don't have transitivity. The concept of chain does not depend on transitivity. In fact the intuitive idea of a chain is without transitivity.

We say that there is a chain between an event a and an event z, if there is a collection of events between a and z related by the relation of causality: $a \hookrightarrow b$, $b \hookrightarrow c, ..., y \hookrightarrow z$. And we adopt the following notation $a \longmapsto z$.

One may want to defend the idea that everything is connected, in the sense that given two events a and b, there is a chain of causality leading from a to b or from b to a. This can be formulated with the following axiom:

$\forall x \forall x ((x \longmapsto y) \vee (x \longleftarrow y))$

4 Case study: citation as a cause-effect phenomenon

We consider here the following possible interpretation of the causality relation: $a \hookrightarrow b$ iff a is cited in b, where a and b are research papers. Research is an important activity of human beings and has been focused more and more on papers. So this example is relevant and up to date.

One may think this is strange, first by wondering if "papers" can be considered as events. What is a paper? We don't pretend here to study in details this question. But our analyzis may shed a side understanding of what a research paper is. First of all a research paper is not a material object, although it may manifest as such. We can roughly say that it is the expression of a research work. There is quite a variation surrounding this "object": a talk given at a conference, a powerpoint file, a preliminary version of the paper circulated among colleagues and friends, the galley-proofs of the paper, the on-line first version, already with an DOI identity number, and finally the "official" paper.

Considering as events only research papers, the relation of causality will be here only among papers. We do not deny that a paper may have been caused by the fly of a stork in a beautiful orange sky, or less poetically by a running board, to recall Poincaré's famous mystical experience [14]. But we consider only one aspect of reality. If a paper a is cited in another paper b, we may imagine that the work of (the authors of) a had some effect on the work of (the authors of) b, that in some sense it is a cause of b.

What are the properties of this causal relation? Let us present the following list:

- it obeys the principle of causality
- it does not obey the reverse principle of causality
- it is non-reflexive
- it is not antisymmetric
- it is not transitive
- it does not obey connectivity

Here are some comments. Principle of causality: a paper citing no other papers would be nowadays really weird. No reverse causality: many papers are not cited, we hope this will not be the case of the present one, that it will have many effects. Non-reflexivity: no comments. No antisymmetry: two papers are cited by each other, this happens and is explained by the process of production / edition of papers. No transitivity: that's reality! No connectivity: connectivity may appear when we restrict the field of research papers, for example considering papers on possible worlds.

Let us make now two remarks about conceptualization. We have here a good example where the notion of chain makes perfectly sense without transitivity. If several papers are cited in one given paper p, this does not mean that these papers have the same effect, at best we can say they have a common effect, p can be see rather as the "product" of these papers.

Acknowledgments and Personal Recollections

Many thanks to my students of the Federal University of Brazil, especially Edson Vinícius Bezerra, who at some point chose causality as a research topic. Thanks also to Claudia Passos, Manuel Mouteira, Manuel Doria, Guilherme Schettini. This work has been concluded when I was visiting scholar at the department of philosophy of the University of California, San Diego, invited by Gila Sher and supported by a CAPES grant (BEX 2408/14-07), my third visit in California after a first visit, at UCLA in 1994, invited by Herb Enderton, and a second visit, at Stanford in 2000 and 2001 invited by Pat Suppes. I take this opportunity to present here some personal recollections about him - see also my recent autobiography [4].

I had been in contact with the ideas of Pat Suppes since a couple of years through his South American connection (Newton da Costa, Rolando Chuaqui, Francisco Doria) before I personally met Pat in Stanford in early 2000 where I was to stay about two years. After that I met him again in several places in the world: in Brazil, France, Switzerland. I organized jointly with Décio Krause a conference for his 80th birthday in Florianópolis in 2002. On the way to this magic island, he stopped in Rio de Janeiro were I was living at the time and we had a nice dinner at the Copacabana Palace with Acacio de Barros and other friends. In Paris we had lunch together with Anne Fagot-Largeault, who had worked with him at Stanford in the 1960s. I remember that we savored sardines directly in their cans in a restaurant on Boulevard Saint-Germain, much in the spirit of the French nouvelle cuisine. In 2004 Pat came to Bern, at the time I was working in the nearby town of Neuchâtel, to receive the Lauener prize, being the first recipient of this prize. We had a good time at this ceremony nicely organized by the manager of the Lauener foundation, Michael Frauchiger, with music and a high quality dinner.

Suppes was in fact a true *bon vivant*, enjoying good food and wine, cars, women, music, sport ... And he was doing research just because this is what he liked the most. When I was at Stanford, he was coming to Ventura Hall nearly every day, conducting seminars, directing the EPGY (Education Program for Gifted Youth) he had created and developing his Brain Lab. He was animating a group of people coming from all over the world.

I was working at the time on many different topics related to logic (universal logic, paraconsistent logic, modal logic, philosophy and history of logic). I presented at the seminar he was organizing with Dagfinn Føllesdal, a criticism of the rejection of propositions by Quine based on Suppes's congruency approach, see [2]. At some point he told me we should write a paper together. His proposal was to discuss and develop with me a general theoretico-philosophical perspective related to the experiments he was conducting on the brain with Marcos Perreau-Guimarães, a Brazilian-French guy he has just engaged. We had daily discussions on the basis on which I started to write my joint paper with Pat, "Semantic computation of truth based on associations already learned" [17]. We tried to develop in this paper a general framework for understanding the relations between language, reasoning and the brain. What is central is the description of a process of association using the axiomatic method. This illustrates a good lesson I learned from Pat Suppes: not to get stuck in some tricky details, but to focus one some fundamental general ideas, motivated by some deep philosophical reflections, developed in an abstract way, nonetheless always connected with reality.

BIBLIOGRAPHY

[1] H.Beebee, C.Hitchcock and P.Menzies, *The Oxford handbook of causation*, Oxford University Press, Oxford, 2009.

[2] J.-Y.Beziau, "Sentence, proposition and identity", *Synthese*, **154**, 2007, pp.371-382.

[3] J.-Y.Beziau, "Badiou et les modèles", in I.Vodoz et F.Tarby (eds), *Autour d'Alain Badiou*, Germina, Paris, 2011.

[4] J.-Y.Beziau, "Logical Autobiography 50", in A.Koslov and A.Buchsbaum (eds), *The Road to Universal Logic Festschrift for the 50th Birthday of Jean-Yves Béziau, Volume II*, Birkhäuser, Basel, 2015.

[5] J.-Y.Beziau and M.V.Kritz, "Théorie et Modèle I: Point de vue général et abstrait", *Cadernos UFS de Filosofia*, **6**, 2010, 9–17.
[6] R.Blanché, *Structures Intellectuelles - Essai sur l'organisation systématique des concepts*, Vrin, Paris, 1966.
[7] D.Bohm, *Causality and chance in modern physics*, Harper, New York, 1961.
[8] R.Chuaqui and P.Suppes, "Free-variable axiomatic foundations of infinitesimal analysis: A fragment with finitary consistency proof", *The Journal of Symbolic Logic*, **60**, 1995, 122–159.
[9] K.Gödel, "An example of a new type of cosmological solution of Einstein's field equations of gravitation", *Reviews of Modern Physics*, **21**, 1949, pp.447-450.
[10] W.Hodges, "Truth in a structure", *Proceedings of the Aristotelian Society*, **86**, 1985-86, pp.135-151.
[11] L.A.Paul and N.Hall, *Causation*, Oxford University Press, Oxford, 2013.
[12] N.Mathieu and A.-F.Schmid (eds), *Modélisation et Interdisciplinarité - Six Disciplines en Quête d'Epistémologie*, Quae, Versailles, 2014.
[13] J.Pearle, *Causality: Models, Reasoning, and Inference*, Cambridge University Press, Cambridge, 2000.
[14] H.Poincaré, *Science et méthode*, Flammarion, Paris, 1908.
[15] B.Russell, "On the Notion of Cause", *Proceedings of the Aristotelian Society*, **13**, 1913, pp.1-26.
[16] A.Schopenhauer, *Über die vierfache Wurzel des Satzes vom zureichenden Grunde, (On the fourfold root of the principle of sufficient reason)*, Rudolstadt, 1813
[17] P.Suppes and J.-Y.Beziau, "Semantic computation of truth based on associations already learned", *Journal of Applied Logic*, **2**, 2004, pp.457-467.
[18] A.Tarski, "The semantic conception of truth and the foundations of semantics", *Philosophy and Phenomenological Research*, **4**, 1944, pp.341-376.

Reconditioning the Conditional
DAVID MILLER

ABSTRACT. Many authors have hoped to understand the indicative conditional construction in everyday language by means of what are usually called conditional probabilities. Other authors have hoped to make sense of conditional probabilities in terms of the absolute probabilities of conditional statements. Although all such hopes were disappointed by the triviality theorems of [15], there have been copious subsequent attempts both to rescue CCCP (*the conditional construal of conditional probability*) and to extend and to intensify the arguments against it. In this paper it will be shown that triviality is avoidable if the probability function is replaced by an alternative generalization of the deducibility relation, the measure of *deductive dependence* of [19]. It will be suggested further that this alternative way of orchestrating conditionals is nicely in harmony with the test proposed in [29], and also with the idea that it is not the truth value of a conditional statement that is of primary concern but its assertability or acceptability.

0   A Critical Memorial to Patrick Suppes

Twenty years ago Karl Popper and I marked Patrick Suppes's 70th birthday with a technical paper [28] that was quite in sympathy with his view of probability as 'perhaps the single most important concept in the philosophy of science' ([35], p. 14). The present tribute, however, though written in gratitude and appreciation, respectfully breaks step. In open disagreement with Suppes's thesis that '[t]he theory of rationality is intrinsically probabilistic in character' ([36], p. 10), I shall sketch, and illustrate the fertility of, a fundamentally non-probabilistic way in which deductive dereliction can be accommodated in the theory of rationality. In short, I shall take exception, not to Suppes's *probabilistic metaphysics*, his view, with which I largely agree, that '[t]he

---

© D. W. Miller 2015. The main idea of this paper (that some indicative conditionals are better understood in terms of deductive dependence than in terms of probability) was mentioned during my presentation 'On Deductive Dependence' at the meeting UNCERTAINTY: REASONING ABOUT PROBABILITY AND VAGUENESS held at the Academy of Sciences of the Czech Republic in September 2006. The details were worked out during a visit to the University of Sassari in the spring of 2013. Warm thanks are due to my Sardinian audience, and also to Alan Hájek and Richard Bradley, who commented on an earlier version of the paper.

fundamental laws of natural phenomena are essentially probabilistic rather than deterministic in character' (*ibidem*), but to his *probabilistic epistemology*. Rejection of a probabilistic approach to rationality is of course to be expected of an adherent of deductivism ([22]). I hold, indeed, that the speculative character of our knowledge can be neither palliated nor controlled by the introduction of probabilities, although its worth may be augmented by sustained criticism. In this paper, however, the thesis to be advanced is a less radical one: that the proposed relaxation of deductive austerity better ministers to the purposes of traditional justificationist epistemology than does an approach that uses probabilities in its management. Whether rationality in any way involves justification will not be examined here.

1 Degrees of Deducibility

Since the time of [5], if not earlier, it has been appreciated that, when p is a probability measure, the identity $p(c \mid a) = 1$ is a necessary, but generally insufficient, condition for the deducibility in classical logic of the conclusion c from the assumption(s) a. What has been less often recognized is that there are other legitimate ways in which *degrees of deducibility* may be measured. In particular, since c is deducible from a if and only if a' is deducible from c' (here the prime stands for negation), the identity $p(a' \mid c') = 1$, which is not equivalent to $p(c \mid a) = 1$, also gives a necessary condition for the deducibility of c from a. There are a number of other interesting possibilities, which I shall elaborate on elsewhere, but they are not the concern of this paper.

A few historical remarks about the function $q(c \mid a) = p(a' \mid c')$ are offered in §8 below. Following [19], §1, we shall call $q(c \mid a)$ the *(degree of) deductive dependence* of the statement c on the statement a, where c is typically the conclusion of an inference from the assumption(s) or premise(s) a. Although, as just noted, $q(c \mid a)$, like $p(c \mid a)$, equals 1 when c is deducible from a, the two functions take the value 0 in different circumstances. Whereas $p(c \mid a) = 0$ when c' is deducible from a, that is, when a and c are mutual *contraries*, $q(c \mid a) = 0$ when c is deducible from a', that is, when a and c are mutual *subcontraries*. In other words, $q(c \mid a)$ assumes the value 1 when c is deductively wholly dependent on a, in the sense of being deducible from a, and the value 0 when c is deductively wholly independent of a, in the sense of having only tautological consequences in common with a. (This relation of deductive independence is closely related to *maximal independence*, as defined by [32].) The interpretation of the function q as a measure of deductive dependence is encouraged by the fact that, if the familiar function $1 - p(b) = p(b')$ is adopted as a measure of the (informative) *content* $ct(b)$ of the statement b, and if $ct(c) \neq 0$, then $q(c \mid a)$ is equal to $ct(c \vee a)/ct(c)$, the 'proportion' of the content of c that resides within the content of a ([13], p. 110; [19], *ibidem*; [17], Chapter 10.4c).

Although the deductive dependence function q has been defined above in terms of the probability function p, this is not supposed to attribute to p any conceptual priority. A more correct treatment would begin with an abstract measure m, and define each of p and q from m. But we forgo such niceties here.

2 Formalities

The function p is required to satisfy the axiom system of [24], appendix ∗v, which is based on the operations of negation ′ and conjunction (inconspicuously represented by concatenation). A dual axiomatic system for the function q, based on the operations ′ and ∨, is presented in [19], §2. In these systems the terms $p(c \mid a)$ and $q(c \mid a)$ are well defined for every a, c, including the contradiction $\bot$ and the tautology $\top$. Indeed, $p(c \mid \bot) = 1 = q(\top \mid a)$ for every a and c. The usual addition or complementation law therefore fails in general, since $p(c \mid \bot) + p(c' \mid \bot) = 2$. But it holds when the second argument of p is not the contradiction $\bot$. Other theorems of the systems will be cited, without much proof, when they are needed. In interpreting Popper's system it is safe to restrict attention to functions p for which $\forall b\, p(c \mid b) \geq p(a \mid b)$ if and only if c is deducible from a. (Since c is deducible from a if and only if a' is deducible from c', the deducibility of c from a can evidently be characterized also by $\forall b\, q(b \mid c) \leq q(b \mid a)$.) It follows that a and c are interdeducible if and only if they are *probabilistically indistinguishable*: that is, $\forall b\, p(c \mid b) = p(a \mid b)$. It should be recorded also that, although $p(c \mid a) = 1$ is in general insufficient for c to be deducible from a, the formula $\forall b\, p(c \mid ab) = 1$ (whose equivalence to the formula $\forall b\, p(c \mid b) \geq p(a \mid b)$ is easily demonstrated within Popper's system[1]) is both necessary and sufficient for deducibility, as is the formula $\forall b\, p(a' \mid c'b) = 1$. In other words, c is deducible from a if and only if $\forall b\, q(b \to c \mid a) = 1$, where the arrow $\to$ represents the material conditional.

3 Conditionals

The appearance here of the material conditional $b \to c$ in the first argument of q may quicken the hope that the substitution of the function q for the probability function p can in some way shed light on the *problem of indicative conditionals*, one of the most tenaciously unsolved problems of modern philosophical logic, and especially on the hypothesis of the *conditional construal of conditional probability* (facetiously dubbed CCCP by [11]). It is the objective of this paper substantially to consummate this hope. But it should be said at once that the matter is not entirely straightforward. Pretty well the simplest form of the CCCP hypothesis worth attending to may be written as the universal identity $\forall a \forall c \forall b\, p(a \rightsquigarrow c \mid b) = p(c \mid ab)$, according to which the absolute probability of the indicative conditional *if a then c* in ordinary language, here shortened to $a \rightsquigarrow c$, is equal to the conditional probability of c given a, not only under the measure p but under any measure obtained from p by *conditionalization* on the statement b. We shall see below that this form of the CCCP hypothesis can hold only for the material conditional $\to$, and that when it does hold, the function p is necessarily two-valued, and no more than a distribution of truth values ([14]). But the identity $\forall a \forall c \forall b\, q(a \to c \mid b) = q(c \mid ab)$, its analogue in

---

[1] If $p(c \mid b) \geq p(a \mid b)$ for every b, then $p(c \mid ab) \geq p(a \mid ab)$. The latter term equals 1, which is the upper bound of the function p. It follows that $p(c \mid ab) = 1$. For the converse we may note that, if $p(c \mid ab) = 1$ for every b, then, by the monotony law for the first argument of p and the general multiplication law, $p(c \mid b) \geq p(ca \mid b) = p(c \mid ab)p(a \mid b) = p(a \mid b)$ for every b.

terms of deductive dependence, may be shown to be equivalent to the CCCP hypothesis, and so to force q to be two-valued too.[2] Moving from p to q in this way does little to avoid triviality.

This result notwithstanding, it is the material conditional $a \to c$ that will be rehabilitated, in §6 below, in terms of the deductive dependence function q.

A great deal has been written on various versions of the CCCP hypothesis and, in particular, on the crucial results of [15] that show that, in the usual Kolmogorov axiomatizations of probability, the hypothesis is condemned in one way or another to triviality. In §4 below it will be shown that, within Popper's axiom system, the triviality of the CCCP hypothesis follows from a result in [25] that is closely related to the theorems of [27]. I shall not discuss directly the implosion of the CCCP hypothesis in Kolmogorov's systems. Nor shall I attempt to summarize the many extensions to Lewis's results and the many responses that have been made to them. For a useful (if dated) discussion, the reader may consult [11], and other papers in the same volume [9], including [31]; and for surveys of the principal philosophical and technical problems posed by conditionals, [8], [4], and the works cited therein. Mention should be made also of [21], which deepens and corrects the theory of tri-events propounded in [6].

4 Triviality of the CCCP hypothesis

In order visibly not to prejudge the question of whether the connective $\leadsto$ introduced above is or is not worthy of the title of an indicative conditional, in this section we shall state the CCCP hypothesis in the ostensibly weaker form

$$\text{CCCP}_0 \qquad \forall a \forall c \exists y \forall b \; p(y \mid b) = p(c \mid ab).$$

We shall show that within Popper's axiomatic system this universal hypothesis implies that for each a, c, the object y can only be the material conditional $a \to c$ and, furthermore, that the values of the function p can only be 0 and 1.

We assume that b is not the contradiction $\bot$. Using a version of the addition law, then the multiplication law, and finally CCCP_0 twice, we may then derive

$$\begin{aligned}
p(ya' \mid b) &= p(y \mid b) - p(ya \mid b) \\
&= p(y \mid b) - p(y \mid ab)p(a \mid b) \\
&= p(c \mid ab) - p(c \mid a(ab))p(a \mid b) \\
&= p(c \mid ab)(1 - p(a \mid b)).
\end{aligned}$$

---

[2]By the definition of q, the identities $q(a \to c \mid b) = q(c \mid ab)$ and $p(b' \mid ac') = p(a' \vee b' \mid c')$ are equivalent. The hypothesis in question therefore holds if and only if $\forall a \forall c \forall b \; p(b' \mid ac') = p(a' \vee b' \mid c')$. By simultaneously replacing in this expression a by b, b by c', and c by a', suppressing the double negations that materialize, and massaging the quantifiers, we obtain $\forall a \forall c \forall b \; p(c \mid ba) = p(b' \vee c \mid a)$. By interchanging a and b, and writing $a \to c$ for $a' \vee c$, we reach $\forall a \forall c \forall b \; p(c \mid ab) = p(a \to c \mid b)$, and finally the CCCP hypothesis for $\to$, as announced.

Using the multiplication law, CCCP$_0$, and the law $p(c \mid \bot) = 1$ we may derive

$$p(ya' \mid b) = p(y \mid a'b)p(a' \mid b)$$
$$= p(c \mid a(a'b))p(a' \mid b)$$
$$= 1 - p(a \mid b),$$

by a second use of the addition law (which is valid here since b is not $\bot$). It follows that if $b \neq \bot$ then $p(c \mid ab)(1 - p(a \mid b)) = 1 - p(a \mid b)$ for all a, c, and hence that $(1 - p(c \mid ab))(1 - p(a \mid b)) = 0$ for all a, c. Now formula (22) in Addendum 3 of [25] states without proof (and in different notation) that $(1 - p(c \mid a))(1 - p(a))$ is equal to the value of the arithmetical difference between the probability $p(a \to c)$ and the probability $p(c \mid a)$. It may be shown more generally that $p(a \to c \mid b) - p(c \mid ab) = (1 - p(c \mid ab))(1 - p(a \mid b))$ when $ab \neq \bot$,[3] which implies that $p(a \to c \mid b) - p(c \mid ab) = 0$ when $ab \neq \bot$. But $ab \equiv \bot$ implies the deducibility of $a \to c$ from b, and hence that $p(a \to c \mid b) = 1 = p(c \mid ab)$. We conclude that $p(a \to c \mid b) - p(c \mid ab) = 0$ for every a, b, c.

It follows from CCCP$_0$ above that for all a, c, there exists a statement y such that $p(a \to c \mid b) - p(y \mid b) = 0$ holds for all b. What this means is that the statement y is probabilistically indistinguishable from the material conditional $a \to c$, in the sense of §2 above, and thus interdeducible with it. The equation $p(y \mid b) = p(c \mid ab)$ can hold for every b if and only if y is the statement $a \to c$.

To show that the function $p(c \mid a)$ can take only the values 0 and 1, we may set aside the case of inconsistent a (since $p(c \mid \bot)$ always equals 1). We have proved above that if $b \neq \bot$ then $(1 - p(c \mid ab))(1 - p(a \mid b)) = 0$, from which it follows that if $p(a \mid b) \neq 1$ then $p(c \mid ab) = 1$ for every c. In particular, $p(a' \mid ab) = 1$. But $p(a'a \mid b) = 0$ if $b \neq \bot$, and so by the multiplication law, $p(a' \mid ab)p(a \mid b) = 0$. It may be concluded that if $p(a \mid b) \neq 1$ then $p(a \mid b) = 0$.

What is so damaging about these results is not that the only conditional conforming to the CCCP hypothesis is the familiar material conditional, for several authors have held that indicative conditionals are, in their semantics, material conditionals, but that all probabilities have to be either 0 or 1. There is nothing but disappointment for the hope that since 'the abstract calculus [of probability] is a relatively well defined and well established mathematical theory ... [and i]n contrast, there is little agreement about the logic of conditional sentences ... [p]robability theory could be a source of insight into [their] formal structure' ([34], p. 64). Indeed, the recourse to probability is otiose, since a two-valued probability function is no more than an assignment of truth values: we may define b to be *true* if $p(b \mid \top) = 1$, and *false* if $p(b \mid \top) = 0$. Matters are actually worse than this, for all true statements turn out to be probabilistically indistinguishable from $\top$, and all false statements probabilistically

---

[3] The right-hand side of the equation, $(1 - p(c \mid ab))(1 - p(a \mid b))$, can be expanded, and by the multiplication law shown equal to $1 - p(c \mid ab) - p(a \mid b) + p(ac \mid b)$. By two applications of the addition law, this can be shown equal to $1 - p(c \mid ab) - p(ac' \mid b) = p(a \to c \mid b) - p(c \mid ab)$.

indistinguishable from $\bot$. This belies the assumption of §2 that probabilistic indistinguishability ought to coincide with interdeducibility.[4]

The first proof that, in Popper's system, $CCCP_0$ implies the two-valuedness of p was given by [14]. The present proof dates from about 1992. The Basic Triviality Result of [20], pp. 301f., which is derivable in Kolmogorov's less general (finite) system, is related but less general.

5 Updating and Relativization

One of the factors that has made the CCCP hypothesis attractive is surely the multiple uses of the word *conditional* and its cognates. As [11] put it, the hypothesis 'sounds right' (p. 80). What is not always realized, however, is that, aside from the word *conditional* in logic, here endorsed, there are two distinct uses of the words in probability theory. There is the process of (Bayesian) *conditionalization*, the generally agreed way in which a probability distribution is updated on the receipt of new information or new knowledge. There is also the result of applying the probability functor p not to a single argument (in the present paper, a statement) but to two arguments, or to one statement relative to another, yielding a binary measure $p(c \mid a)$ that is standardly called *conditional probability*. These processes of *updating* and *relativization*, as they will hereafter be called, happen to have the same mathematical effect: the result of updating the singular measure p with the information b is the same as relativizing it to b. It follows that updating $p(c)$ with b, and then relativizing it to a, is the same as relativizing $p(c)$ to a, and then updating it with b. Since conjunction in the second argument of p is commutative, the outcomes $p(c \mid ba)$ and $p(c \mid ab)$ are identical. Although relativization and updating are therefore formally dead ringers for each other, they deserve to be understood as distinct undertakings. In particular, if $p(c \mid a) = r$ is a declaration of relative probability there is no presumption that the statement a is known to be true, or even supposed to be true ([37], §2), any more than this is the case in the metalogical declaration $a \vdash c$. (But the interpretation of a as a statement of evidence, and of c as a hypothesis, is not excluded.) This is not idle pedantry. With the function q, the distinction between updating and relativization emerges as a distinction with a difference.

The axiomatic system of [24] that we adopted in §2 above is a system of relative probability $p(c \mid a)$. It is easy to check that if the function p satisfies the axioms, and if $b \neq \bot$, then $p_b(a \mid c) = p(a \mid cb)$ also satisfies them. (The function $p_\bot$ is identically equal to 1, and violates the axiom that requires the function p to have at least two distinct values.) The subscript notation embodied in p_b will be used whenever we wish to refer to the updating of a function with the information b. Since $p_b(c \mid a)$ equals $p(c \mid ab)$ for every a,

---

[4]The two-valuedness of p settles the truth table for negation. The other tables need also the addition and monotony laws. For example, by the general addition law, $p(a \to c \mid \top) = 0$ if and only if $p(a \mid \top) = 1 - p(ac \mid \top)$. By monotony and two-valuedness, this holds if and only if $p(a \mid \top) = 1$ and $p(ac \mid \top) = 0$. In short, $a \to c$ is false if and only if a is true and c is false. The CCCP hypothesis implies that in addition $a \to c$ is false if and only if $p(c \mid a) = 0$. But if c is true, $a \to c$ is true for every a, and accordingly $p(c \mid a) = 1 = p(\top \mid a)$ for every a.

and hence $p_b(b \mid a) = p(b \mid ab) = 1 = p(\top \mid ab) = p_b(\top \mid a)$, updating with b amounts to a decision to treat b as probabilistically indistinguishable from $\top$.

Since $q(c \mid a) = p(a' \mid c')$, the updated function q_b is defined by $q_b(c \mid a) = p_b(a' \mid c') = p(a' \mid c'b) = q((c'b)' \mid a)$, which equals $q(b \to c \mid a)$. In general, this term differs from $q(c \mid ab)$. Updating with b is not the same as relativizing to b. The distinction is especially transparent when the second argument of the function q is the tautology $\top$. For except when $a \equiv \bot$, the value of $p(\bot \mid a)$ is 0 for every probability measure; and therefore $q(c \mid \top) = 0$ except when $c \equiv \top$. (The function q, unlike the function p, has an almost flat prior distribution.) Updating p to p_b does not change matters: $q_b(c \mid \top)$ still equals 0 (unless $c \equiv \top$). But relativization of $q(c)$ to b yields $q(c \mid b)$, which may well not be 0.

6 The Reconditioned Conditional

Armed with these considerations we are at last in a position to understand how and why the replacement in the CCCP hypothesis of the probability measure p by the deductive dependence measure q makes such a dramatic difference. The first formula displayed below is $CCCP_0$, exactly as it was displayed in §4. The formula $CCCP_1$ is a notational variant, obtained from $CCCP_0$ by writing $p_b(c \mid a)$ for $p(c \mid ab)$. The formula $CCCP_2$ is obtained from $CCCP_0$ by first commuting the terms in the conjunction ab, then interchanging the letters a and b throughout, and finally writing $p_b(c \mid a)$ for $p(c \mid ab)$, as before. It is because updating and relativization are formally equivalent manoeuvres that each of $CCCP_1$ and $CCCP_2$ is equivalent to $CCCP_0$, though they look different.

$CCCP_0$ $\qquad\qquad \forall a \lor c \exists y \forall b \; p(y \mid b) = p(c \mid ab)$
$CCCP_1$ $\qquad\qquad \forall a \lor c \exists y \forall b \; p(y \mid b) = p_b(c \mid a)$
$CCCP_2$ $\qquad\qquad \forall b \lor c \exists y \forall a \; p(y \mid a) = p_b(c \mid a)$.

We now replace p by q in both $CCCP_1$ and $CCCP_2$, to produce the formulas

$CCCQ_1$ $\qquad\qquad \forall a \lor c \exists y \forall b \; q(y \mid b) = q_b(c \mid a)$
$CCCQ_2$ $\qquad\qquad \forall b \lor c \exists y \forall a \; q(y \mid a) = q_b(c \mid a)$.

These formulas are far from equivalent to each other: one is refutable, the other is demonstrable. $CCCQ_1$ is refuted by identifying b with $\top$. This shows that, for each a and c, $q(c \mid a) = q_\top(c \mid a)$ can take only the value 1 or the value 0; the value 1 if y (which may depend on a and c) is equivalent to $\top$, and the value 0 if it is not. In contrast, $CCCQ_2$ is demonstrable, since y may be the conditional $b \to c$. As was shown near the end of §5 above, $\forall b \forall c \forall a \; q(b \to c \mid a) = q_b(c \mid a)$.

7 Discussion

In the interests of amity and brevity, I shall limit my discussion of these results to three items. One concerns their relation to the well-known Ramsey test. A second concerns the tenability of the thesis that, at least with regard to conditionals, measures of deductive dependence offer an attractive alternative to measures of probability. The third matter, dealt with first, and in only a couple of sentences, is whether the unassailability of $CCCQ_2$ vindicates the identification of all indicative conditionals, at a semantic level, with material conditionals. This remains an open question. But I am not able here to provide solace to those who, having resolved to learn about indicative conditionals by studying their synergy with probabilities, are dismayed by what has been learnt.

Ramsey's test Much work on the connection between conditionals and probability has been guided by the words of Ramsey in 1929 ([29], p. 247): 'If two people are arguing "If p, will q?" and are both in doubt as to p, they are adding p hypothetically to their stock of knowledge, and arguing on that basis about q; ... We can say that they are fixing their degrees of belief in q given p. If p turns out false, these degrees of belief are rendered *void*.' In [33], p. 101, this description becomes a piece of advice: 'your deliberation ... should consist of a simple thought experiment: add the antecedent (hypothetically) to your stock of knowledge (or beliefs), and then consider whether or not the consequent is true. Your belief about the conditional should be the same as your hypothetical belief, under this condition, about the consequent.' [11], p. 80, add that the agent's system of beliefs may need to be revised (but as little as possible) if it is to accommodate the antecedent consistently, a qualification that imports new problems. What lies behind the advice, if I understand it, is the idea that evaluating the probability of the consequent of a conditional, relative to its antecedent, is a way in which the agent might 'consider whether or not the consequent is true'.

I suggest that the explicit identity that we may extract from $CCCQ_2$, namely $q(b \to c \,|\, a) = q_b(c \,|\, a)$, heeds this advice as well as does any identity derivable from the CCCP hypothesis. To be sure, there is a difference. In the case of an identity of the form $p(a \rightsquigarrow c \,|\, b) = p(c \,|\, ab)$, it is likely that what Stalnaker (and others) had in mind was that the antecedent of the conditional $a \rightsquigarrow c$ be 'added to your stock of knowledge (or beliefs)' by further relativizing $p(c \,|\, b)$ to a. I do not know that this strategy has ever been described (equivalently) as one of updating of $p(c \,|\, b)$ with a. But in the identity $q(b \to c \,|\, a) = q_b(c \,|\, a)$, the antecedent of the conditional $b \to c$ is unambiguously used to update the function q. This is how b is to be 'added to your stock of knowledge (or beliefs)'.

Stated quite literally, what is here being proposed is this: in order to assess the deductive dependence of the material conditional $b \to c$ on the statement a, the agent should (provisionally and hypothetically) update the function q to q_b and then, using this updated function, assess the deductive dependence of c on a. This procedure cannot properly be described as 'evaluating the dependence of the consequent of a conditional on its antecedent'. But if a is supposed to state truthfully some information about the world, it is surely one way in which the agent might 'consider whether or not the consequent is true'.

Assertability and Acceptability of Conditionals It has been suggested by several writers, especially [1], that conditionals cannot be true or false, and that $p(c\,|\,a)$ measures not the probability of the truth of $a \rightsquigarrow c$, but its *assertability*; that is to to say, the appropriateness of its utterance. Others, including Adams himself in a later phase ([2]), have favoured the term *acceptability*, that is to say, the reasonableness of the belief in $a \rightsquigarrow c$. [10], §2, has ventured the neologism *assentability*. Although this has to my ears a subjectivist ring that is absent from *acceptability* and, to a lesser extent, *assertability*, for our present purposes the differences between these ideas are less important than what they have in common, which is an origin in the justificationist doctrine that an agent is entitled fully to assert or to accept or to assent to a statement only if he knows it to be true. The word *probably*, and similar expressions such as *in my opinion* and *I think*, are often used to qualify statements that are not fully asserted. The less probable that c is, given a, the less the agent is entitled to assert it, or the more tentatively he asserts it. In this vein, [16], Chapter 1, called probability 'a guarded guide'.

Those of us who dismiss as not quite serious the goal of justified truth never worry that we are not entitled to assert a statement. We think that we are entitled to say what we like, whatever the epistemological authorities may enjoin. But we may worry whether a proposition asserted is true, and if we suspect that it is not, we may qualify our assertion by such expressions as *about* or *or so* or *roughly* or *more or less*. Since the quantity $q(c\,|\,a)$, the deductive dependence of a non-tautological statement c on a statement a, is a straightforward measure of how well (the content of) c is approximated by (the content of) a, ranging from 0, when a contains none of c, to 1 when it contains it all, it does appear that $q(c\,|\,a)$ may serve also as a measure of the assertability or the acceptability of the statement c in the presence of a. If our aim is truth, then the higher $q(c\,|\,a)$ is, the more successful is the statement (or hypothesis c), given the statement (or evidence) a. More generally, the assertability or acceptability of the conditional $b \rightarrow c$ may be measured by $q(b \rightarrow c\,|\,a)$, that is, by $q_b(c\,|\,a)$. It is vigorously denied here that the 'highly entrenched tenet of probabilistic semantics ... [that] the assertability of conditionals goes by conditional probability' ([3], p. 584) exhausts the senses in which a conditional statement may be assertable or acceptable, but not completely so.

8 Conclusion

The goal of this paper has been to elucidate one of the gains that can be made in epistemology by replacing probability measures (understood as degrees of belief) by measures of deductive dependence (understood as degrees of approximation). On this theme, much more needs to be said than can be said here. In the first place, it must be recognized that variants of the function q of deductive dependence have been introduced before, in rather different contexts. [12], Part IV, for example, interpreted $q(a\,|\,c)$ as a measure of the *systematic power* of the hypothesis c to organize the evidence a. [30], appendix, espied in the divergence between the functions p and q a potential solution to Hempel's paradoxes of confirmation. [13], §IV, interpreted $q(a\,|\,c)$ as a measure of the *information transmitted* by the evidence a about the hypothesis c, and used it to answer Ayer's question of why those who assay hypotheses by their relative probabilities ever search for new evidence. The function q has similarities also with the idea of probabilistic validity advanced in [2], and especially with the use of p-values in modern classical (non-Bayesian) statistics. All these connections will have to be explored in due course. Interested readers may glean from [18] meanwhile a glimpse of the versatility of the function q, and of the role that it may perform in a saner philosophy of knowledge than is fashionable at present.

BIBLIOGRAPHY

[1] Adams, E. W., The Logic of Conditionals. *Inquiry* 8: 166–197, 1965.
[2] Adams, E. W., *A Primer of Probability Logic*. Stanford CA: Center for the Study of Language and Information, 1998.
[3] Arló-Costa, H., Bayesian Epistemology and Epistemic Conditionals: on the Status of the Export–Import Laws. *The Journal of Philosophy* XCVIII: 555–593, 2001.
[4] Arló-Costa, H., The Logic of Conditionals. In: Zalta, Edward N. (ed.) *The Stanford Encyclopedia of Philosophy (Summer 2014 Edition)*, 2014.
[5] Bolzano, B. P. J. N., *Wissenschaftslehre*, 1837. (English translation: *Theory of Science*. New York: Oxford University Press, 2014.) Sulzbach: Seidelsche Buchhandlung.
[6] De Finetti, B., La logique de la probabilité. *Actualités scientifiques et industrielles*: 391, pp. 31–39. (English translation: *The Logic of Probability*, Philosophical Studies 77:181–190, 1995.)
[7] De Finetti, B., The Logic of Probability. English translation of [De Finetti, 1936]. *Philosophical Studies* 77: 181–190, 1995.
[8] Edgington, D. M. D., Conditionals. In: Zalta, Edward N. (ed.) *The Stanford Encyclopedia of Philosophy (Spring 2014 Edition)*, 2014.
[9] Eells, E. and Skyrms, (editors), B. *Probability and Conditionals. Belief Revision and Rational Decision*. Cambridge: Cambridge University Press, 1994.
[10] Hájek, A., The Fall of "Adams' Thesis"? *Journal of Logic, Language and Information*, 21: 145–161, 2012.
[11] Hájek, A. and Hall, N., The Hypothesis of the Conditional Construal of Conditional Probability. In: Eells, E. and Skyrms, B. (eds.) *Probability and Conditionals. Belief Revision and Rational Decision*. Cambridge: Cambridge University Press 75–111, 1994.
[12] Hempel, C. G. and Oppenheim, P., Studies in the Logic of Explanation. *Philosophy of Science* 15: 135–175, 1948.
[13] Hilpinen, I. R. J., On the Information Provided by Observations. In: Hintikka, K. J. J. and Suppes, P. (eds) *Information and Inference*. Dordrecht: D. Reidel Publishing Company, 237–255, 1970.
[14] Leblanc, H. and Roeper, P., Conditionals and Conditional Probabilities. In: Kyburg Jr, H. E. and Loui, R. P. and Carlson, G. N. (eds.) *Knowledge Representation and Defeasible Reasoning*. Dordrecht: Kluwer Academic Publishers, 287–306, 1990.
[15] Lewis, D. K. Probabilities of Conditionals and Conditional Probabilities. *The Philosophical Review* **85** 297–315 1976
[16] Lucas, J. R., *The Concept of Probability*. Oxford: Clarendon Press, 1970.
[17] Miller, D. W., *Critical Rationalism. A Restatement and Defence*. Chicago & La Salle IL: Open Court. 1994.
[18] Miller, D. W., If You Must Do Confirmation Theory, Do It This Way. In: http://www.warwick.ac.uk/go/dwmiller/itam.pdf 2014.
[19] Miller, D. W. and Popper, K. R. Deductive Dependence. In: *Actes IV Congrés Català de Lògica*. Barcelona: Universitat Politècnica de Catalunya & Universitat de Barcelona, pp. 21–29, 1986.
[20] Milne, P., The Simplest Lewis-style Triviality Proof Yet?. *Analysis* 63: 300–303, 2003.
[21] Mura, A. M., Towards a New Logic of Indicative Conditionals. *Logic and Philosophy of Science* IX: 17–31, 2011.
[22] Popper, K. R., *Logik der Forschung*. Vienna: Julius Springer Verlag, 1935.
[23] Popper, K. R., *The Open Society and Its Enemies*. London: George Routledge & Sons, 1945.
[24] Popper, K. R., *The Logic of Scientific Discovery*. Augmented English translation of [Popper, 1935]. London: Hutchinson Educational, 1959.
[25] Popper, K. R., *Conjectures and Refutations. The Growth of Scientific Knowledge*. London: Routledge & Kegan Paul, 1963.
[26] Popper, K. R., *The Open Society and Its Enemies*. 5th edition of [Popper, 1945]. London: Routledge & Kegan Paul, 1966.
[27] Popper, K. R. and Miller, D. W., A Proof of the Impossibility of Inductive Probability. *Nature* **302**: pp. 687f., 1983.

[28] Popper, K. R. and Miller, D. W. Contributions to the Formal Theory of Probability. In: Humphreys, P. W. (ed.) *Patrick Suppes: Scientific Philosopher. Volume I. Probability and Probabilistic Causality*. Dordrecht, Boston MA, & London: Kluwer Academic Publishers, pp. 3–23, 1994.
[29] Ramsey, F. P., General Propositions and Causality (1929). In: Braithwaite, R. B. (ed.) *The Foundations of Mathematics*. London: Routledge & Kegan Paul, pp. 237–255, 1931.
[30] Reichenbach, H., *Nomological Statements and Admissible Operations*. Amsterdam: North-Holland Publishing Company, 1954.
[31] Suppes, P. Some Questions about Adams' Conditionals'. In: Eells, E. and Skyrms, B. (eds) *Probability and Conditionals. Belief Revision and Rational Decision*. Cambridge: Cambridge University Press, 5–11, 1994.
[32] Sheffer, H. M., The General Theory of Notational Relativity. In: Brightman, E. S. (ed.) *Proceedings of the Sixth International Congress of Philosophy*. Cambridge MA: Harvard University Press, 348–351, 1926. (Original mimeograph version, Department of Philosophy, Harvard University, 1921.)
[33] Stalnaker, R. C., A Theory of Conditionals. In: Rescher, N. (ed.) *Studies in Logical Theory. American Philosophical Quarterly Monograph Series, 2*. Oxford: Blackwell, 98–112, 1968.
[34] Stalnaker, R. C., Probability and Conditionals. *Philosophy of Science* 37¿ 64–80, 1970.
[35] Suppes, P., *Representation and Invariance of Scientific Structures*. Stanford CA: Center for the Study of Language and Information, 2002.
[36] Suppes, P., *Probabilistic Metaphysics*. Oxford & New York: Blackwell, 1984.
[37] van Fraassen, B. C., Fine-grained Opinion, Probability, and the Logic of Full Belief. *Journal of Philosophical Logic* 24: 349–377, 1995.

A 6-Valued Calculus which Avoids the Paradoxes of Deontic Logic

PAUL WEINGARTNER

ABSTRACT. he paper offers a 6-valued system of Deontic Logic which is decidable and has the finite model property. It is based on a 6-valued propositional logic RMQ which has relevance-properties and has been developed for avoiding paradoxes in different areas, especially when logic is applied to modern physics (The Review of Symbolic Logic 2009). The present deontic system results from RMQ by adding an operator for "obligatory". The definitions for "permitted" and for "forbidden" are the usual ones. The proposed system avoids all well-known paradoxes of Deontic Logic including the more complicated paradoxes of Chisholm and Forrester.

1 Introduction

It is a great pleasure and honor to contribute to the volume in honor of Pat Suppes' 90th birthday. I know Pat from several international conferences since 1965 and I remember with pleasure his first lectures (1974) in Salzburg on the occasion of a Summer School which I organized. The six-valued logic proposed here deviates from Classical Logic. And Pat has a serious warning against that: "There are, however, a number of reasons for moving very slowly to the adoption of a new logic, especially a logic that is clearly weaker than classical logic" (Suppes (1984) p.91).

However the six-valued systems I developed include all theorems of Classical Propositional Calculus (CPC) as materially (classically) valid. They deviate from CPC only for the subset of their strictly valid theorems, which throw out redundancies and irrelevances of a certain type allowed by CPC. It can be shown that if this subset of strictly valid theorems is applied to empirical sciences many paradoxes in different areas can be avoided (cf. section 2., 7 and 8 below).

The purpose of this paper is to show that the subsequently described 6-valued logic (6-DL) avoids the following well known paradoxes of Deontic Logic: Ross-Paradox, paradoxes of Derived Obligation, of Good Samaritan, of Free Choice, of Commitment, of Forrester, of Chisholm.

The 6-valued logic, 6-DL, results from the 6-valid logic RMQ[1] by adding the deontic operator O (for: obligatory), which turns the 6 values of the basic

---

[1]RMQ has been developed in Weingartner (2009). Cf. further (2010a), (2010b) and

matrix 1 2 3 4 5 6 of the proposition p into the 6 values 3 6 5 1 4 6 of the proposition Op.

2 Motivation

2.1 The underlying system RMQ

The underlying system RMQ was developed with the intention to construct a propositional logic which contains its own semantics and obeys some important criterion of relevance. This relevance criterion was developed together with Gerhard Schurz (Schurz and Weingartner (1987)). in order to avoid different types of paradoxes in the domains of scientific explanation, disposition predicates, scientific confirmation, verisimilitude, Quantum Physics and Deontic Logic. The paradoxes of Deontic Logic is only some small branch of difficulties besides the others mentioned. The most well-known of them (see 1.-5. of section 5) do not seem to be specific since they have a common cause. The common causes are principles of CPC (Classical Propositional Logic) that contain irrelevant elements (such elements that can be replaced by any others or can be reduced to others) in the conclusion or in the consequence class. The relevance criterion has two parts, a replacement part RC and a reduction part RD. The main idea of the first part of the relevance criterion — called replacement criterion (RC) — is to forbid those parts of a consequence (conclusion) of a valid inference which can be replaced (on one or more occurrences) by any arbitrary part (wff) *salva validitate* of the inference.

For example the classically valid principle of addition $p \to (p \vee q)$ allows to introduce a new sentence q which has nothing to do with the premises and can be replaced by any other sentence (wff) *salva validitate* of the inference. It is important to realize that this principle is the chief cause for the following paradoxes in different domains: Hesse's paradox of confirmation, Goodman's paradox, paradox in the definition of verisimilitude, Ross' paradox.

Similarly, the classical valid *ex falso quodlibet* principle $\neg p \to (p \to q)$ is the chief cause for the disposition paradox and also for the paradoxes of Derived Obligation and Commitment.[2] Therefore we do not think that the paradoxes of Deontic Logic are a special type of paradoxes just in this domain. On the contrary, the underlying cause is much more general. It consists of some very tolerant properties of Classical Logic concerning valid inferences which are properties of irrelevance in the sense that something which can be replaced (in the consequence-class) by anything arbitrary cannot be relevant. The second part of the relevance criterion is a reduction criterion (RD) which reduces redundant repetitions, double negations, splits complex wffs into smallest conjuncts. The reduction criterion is however not necessary to be applied in this study for avoiding paradoxes of Deontic Logic.

2.2 More than two truth-values

To construct a propositional logic where the theorems obey the above relevance criterion RC and all classically valid theorems receive a designated value,

---

(2011).

[2]This has been shown in Weingartner (2001).

is of course not possible with 2 values, since this is just 2-valued Classical Logic which violates this criterion. Since we do not want to introduce values which are neither true nor false, like the values "indefinite" in the well-known three-valued logics (Łukasiewicz, Kleene, Bochvar), the option is to have more than one value for truth and more than one value for false. The first option is a four-valued system which obeys to a high degree the criterion RC. The best proposal so far seems to be the four-valued system described in Weingartner (2009), p. 146. It avoids several principles of CPC, like the principle of addition, which violate RC. But it still has a number of drawbacks because it is not differentiated enough: The above mentioned *ex falso quodlibet* principle is strictly valid and therefore the deontic paradoxes of Derived Obligation and of Commitment cannot be avoided. Since Frege?s law $p \to (q \to p)$ holds strictly, several forms of commensurability are valid on logical grounds, which says that this system cannot be applied to modern physics. The classically valid but intuitionistically invalid principle $p \vee (p \to q)$ is strictly valid although it violates RC. This list of disadvantages could be continued. A natural way to achieve more differentiation is the extension to six values. Here a special distribution of values can be found — realized by the system RMQ — that is differentiated enough to approximate very closely the relevance criterion RC and that is able to avoid paradoxes in different domains (from philosophy of science to physics and to Deontic Logic) which come up when CPC is applied. Theoretically, still more differentiation would be possible with eight values. But the question concerning 8 values is that of diminishing returns. If 6 values satisfy already the important desiderata, the extension to 8 values seems to be a kind of empty precision or differentiation. However, one has to be open-minded concerning such an extension: especially if one wants to introduce new independent operators, for example to introduce not only a deontic operator for obligation, but also an independent action operator in order to express statements like "it ought to be the case that A acts in such a way that p occurs" and consequently also different application of negations (not-act that p, act that not-p), then such an extension seems necessary. This can be guessed from the fact that with six values the system is not suitable for iterations (see section 6. below). If a certain independence of new operators is not guaranteed, necessary distinctions are dropped and unacceptable consequences follow. However, such investigations (concerning extension to more than 6 values) have to wait for further research. Concerning odd values (5 or 7 or 9) the disadvantage is that the negation is not symmetric and this causes a lot of anomalies.

2.3 RMQ and its deontic specification 6-DL

RMQ distinguishes two concepts of validity. If the highest value of the matrix (to every wff there corresponds exactly one matrix) called the characteristic value $cv = 3$, then the formula is classically (materially) valid. If the $cv = 2$, the formula is strictly valid. The strictly valid formulas of RMQ form a subset of CPC and possess relevance properties. The strictly valid implicational formulas have many similarities with the restricted implication in Conditional Logic and are therefore especially suitable for interpreting the deontic conditional $O(q/p)$. For example the three options to interpret $O(q/p)$ with an implication as discussed by Chellas (Chellas (1980), p. 201.) do not collapse into tautologies, when interpreted with the strict implication of 6-DL: Under the condition of false antecedents or true consequence, neither of the three options $p \to Oq$, $O(p \to q)$, $\Box(p \to Oq)$ are true; moreover, none of these three is implied by $O\neg p$ or by Oq. $O(p \to q)$ does not imply $O(p \wedge r \to q)$, where r is an arbitrary proposition. As a result, all the three options of Chellas for interpreting the deontic conditional $O(q/p)$ are possible in 6-DL, where $O(p \to q)$ seems to be the most preferable.

It should be observed that formulas invalid in 6-DL, as all the well-known deontic paradoxes (see below section 5.), are of course still invalid if their main implication is strengthened to a necessary or strict implication. Also classically valid (but strictly invalid) formulas become simply invalid by such a strengthening, because classical validity, i.e. $cv = 3$ turns into $cv = 6$.

2.4 Choosing the matrices for RMQ and 6-DL

(1) Choosing the matrix of the underlying system RMQ The first desideratum for choosing the matrix of RMQ was Patrick Suppes' warning cited at the beginning of this paper: The deviation from classical logic (in this case classical propositional calculus CPC) should be as modest as possible. This is satisfied by RMQ in the sense that all theorems of CPC are at least classically (materially) valid ($cv = 3$) in RMQ.

The second desideratum was that the strictly valid ($cv = 2$) formulas of RMQ should satisfy a simple relevance idea which proved satisfactory for solving difficulties in many different domains when logic is applied to empirical sciences (see section 2 points 7 and 8 below).

Observe that the matrices of RMQ are extremely sensitive w.r.t. any change. Already a change in inside truth (values 1, 2, 3) or inside falsity (values 4, 5, 6) has lots of consequences. A change from false to true or from true to false would immediately destroy our first desideratum that all CPC theorems are materially valid in RMQ. This sensitivity is connected with great differentiation: A simpler 6-valued modal system needed 30 axioms to describe (the matrices of) the system. Axiomatisation of RMQ is under investigation.

(2) Choosing the matrix for 6-DL A first desideratum was to build 6-DL on a system of propositional logic with relevance properties which proves satisfactory for solving difficulties when implication is applied to empirical sciences. This means that the inbuilt relevance should already avoid most of the so-called paradoxes of implication. The reason for this is that many of the paradoxes of Deontic Logic arise from the principle of addition and from *ex falso quodlibet* principles: the first five (and most well-known) paradoxes listed in section 5. below are in fact not specific to Deontic Logic, since they result from these two principles which are the chief cause for many paradoxes in different areas (see 1.1 above). Only the Forrester and the Chisholm paradox are specifically deontic.

A second desideratum was to choose the matrix for Op in such a away as to meet the usual definability conditions of Fp and Pp, where all these are contingent propositions being independent of p and $\neg p$.

The third desideratum was that the so established decidable system 6-DL has theorems agreed upon by most of the systems of Deontic Logic. This is shown by the theorems 1-26 of section 4.1 below.

A fourth desideratum was that 6-DL should profit from the fact that the strict implication of RMQ comes close to the restricted implication of Conditional Logic. Principles which do not obey such restrictions should be only materially (classically) valid. Examples for such principles are the materially valid (but strictly invalid) principles of section 4.2. For example, 2. and 6. of 4.2 show that an implication should not follow from a conjunction (though classically valid) if it should express some connection beyond just having designated values. Therefore 2., 4. and 6. are strictly invalid. Similarly, 8., 9. and 10. of 4.2 result from the principle of addition and that of ex falso quodlibet, two classical culprits of several paradoxes in different areas. Also they are strictly invalid in 6-DL. The other principles of 4.2 are also problematic w.r.t. a standard system of Deontic Logic.

2.5 Comparison with other systems of Deontic Logic

The system 6-DL agrees in many theorems with the Standard Deontic Logic (SDL) (see section 4.1). On the other hand, SDL suffers from all paradoxes listed in section 5 below and needs additional restrictions to avoid them; whereas 6-DL avoids all of them. There have been several approaches to avoid the paradoxes. One is that of restricting the classical implication by some system of Conditional Logic. This approach is close to the relevance restrictions concerning the implication in 6-DL. This similarity is shown by the fact that the first 5 well-known paradoxes of deontic logic (cf. section 5) are not specifically deontic, since their common reasons are the too tolerant properties of the implication in Classical Logic; that means that the underlying classical principles lead to difficulties in many different domains of application. And this weakness of classical implication is strengthened by both Conditional Logic and by RMQ and 6-DL.

Concerning particular approaches to avoid the deontic paradoxes — this is not the place to give an overview of the numerous approaches — mainly two ideas have been used: (1) To invent a new system of Deontic Logic. (2) To drop or weaken some rules or to change the symbolic interpretation of claims made in natural language. An example for the first is Meyer?s interpretation of Deontic Logic with the help of Dynamic Logic. This looks promising, although it does not solve the paradoxes of Ross and Commitment. However, the system was refuted as a whole because of an absurd consequence proved by A. Anglberger: Under the two defensible assumptions of (1) substitution of equivalent action-terms and (2) $Op \to Pp$, it can be proved in that system that no possible action is forbidden.[3] A further example of the first idea is Mott?s system of Factual Detachment (1973). However, one disadvantage is that it generally forbids deontic detachment, i.e. the inference from Op and $O(q/p)$ to Oq. In 6-DL, deontic detachment is strictly valid if $O(q/p)$ is interpreted as $O(p \to q)$, but invalid if interpreted as $p \to Oq$ or as $p \Rightarrow Oq$.

---

[3] Anglberger (2008).

Factual detachment (the inference from p and $p \to Oq$ to Oq) is strictly valid in 6-DL as an instance of modus ponens. As a result, both factual detachment and deontic detachment (interpreted with $O(p \to q)$) are not only compatible, but are strictly valid in 6-DL. And Chisholm?s paradox is invalid, i.e. the four premises are contingent, but not contradictory (see section 5 below for details).

An example for dropping or weakening some rules is to drop the principle of inheritance in order to avoid the paradox of the Good Samaritan or the Forrester paradox.[4] Another is to drop deontic detachment to avoid Chisholm?s paradox.[5]

A general criticism to such remedies is that they are all too local. They do not try to meet the problem at the root. As already pointed out, most of the well-known paradoxes of Deontic Logic are not specifically deontic. They are based on too liberal principles of Classical Logic, which lead to difficulties in very different and otherwise unrelated domains as Modern Physics, Philosophy of Science and Deontic Logic. And those that are specifically deontic, are the paradox of Chisholm and that of Forrester. Concerning some versions of them, one can easily see that the conjunction of the premises is contradictory, just by applying modus ponens and the generally agreed distribution of O over $\to$. In this case, one can hardly speak of a real paradox at all. In other versions the premises are only contradictory when some classically valid but irrelevant (according to RMQ) implications or inferences are presupposed. Since 6-DL throws them out as strictly invalid, the conjunction of the premises are no more contradictory and the respective versions of these paradoxes are invalid and no more paradoxical.

3 The System RMQ

The system RMQ is defined as the set of all formulas (wffs) which are satisfied by the matrix $M = \langle T, F, L \rangle$, where $T = \{1,2,3\}$, $F = \{4,5,6\}$ and the operations $\neg, \vee, \wedge, \to$ are defined as follows:

p	$\neg p$	$p \vee q$	1	2	3	4	5	6
1	6	1	1	1	1	1	1	1
2	5	2	1	2	2	2	1	2
3	4	3	1	2	3	1	3	3
4	3	4	1	2	1	4	4	5
5	2	5	1	1	3	4	5	5
6	1	6	1	2	3	5	5	6

The system RMQ was developed in order to avoid paradoxes when logic is applied to physics and to several problems in the philosophy of science. RMQ has the following properties:

---

[4]Cf. Goble (1991). See below section 5, 6c.
[5]Cf. Mott (1973). Mott?s interpretation is 7a of section 5.

$p \wedge q$	1 2 3 4 5 6	$p \to q$	1 2 3 4 5 6	Lp
1	1 2 3 4 5 6	1	1 2 3 5 5 6	1
2	2 2 3 4 6 6	2	1 1 3 5 5 5	3
3	3 3 3 6 5 6	3	1 2 1 4 5 5	6
4	4 4 6 4 5 6	4	1 2 3 1 3 3	6
5	5 6 5 5 5 6	5	1 2 2 2 1 2	6
6	6 6 6 6 6 6	6	1 1 1 1 1 1	6

1. RMQ is a 6-valued matrix system (3 values for truth, 3 for falsity) and so it contains its own semantics. Every well-formed formula of RMQ is unambiguously determined by a particular matrix which contains 6n values for n ($n = 1,2,\ldots$) different propositional values.
2. RMQ is motivated by two relevance criteria called replacement (RC) and reduction (RD), which avoid difficulties in the application of logic (see 6) and 7) below).[6]
3. RMQ is consistent and decidable.
4. RMQ has the finite model property.
5. RMQ has two concepts of validity: a weaker one (classically valid which is identical with materially valid) and a stronger one (strictly valid). All theorems of two-valued Classical Logic (Classical Propositional Calculus CPC) are at least classically valid, that is materially valid, in RMQ. Only a restricted class of them are strictly valid in RMQ.
6. The validity of a proposition is decided by calculating the highest value (cv) in its matrix. If $cv = 3$ the proposition (formula) is classically valid, that is materially valid. If $cv = 2$ the proposition (formula) is strictly valid.
7. The strictly valid theorems of RMQ avoid a great number of well-known paradoxes in the domain of scientific explanation, law statements, disposition predicates, verisimilitude, ... etc.[7]
8. The strictly valid theorems of RMQ avoid the well-known difficulties when logic is applied to physics; especially those with commensurability, distributivity and with Bell's inequalities.[8]
9. RMQ is closed under transitivity of implication, under *modus ponens*, and under equivalence substitution.
10. RMQ also contains a modal system with 14 modalities, where Lp (necessary p) has the matrix: 1 3 6 6 6 6. $Mp = \neg L \neg p$ (possible p).

---

[6]For an exact formulation of these criteria see Weingartner (2009) §2 and (2010b) section 2

[7]See Weingartner (2009) section 4.3, (2010a) section 2.4, (1994).

[8]See Weingartner (2009) sections 2.1, 2.2 and 4.2, 4.4.

4 The Deontic System 6-DL

The deontic logic 6-DL results from RMQ by adding the deontic operator O for *obligatory*. The operator O applied to the proposition p turns the basic matrix of p 1 2 3 4 5 6 into the following matrix for Op: 3 6 5 1 4 6. Since 6-DL obeys the usual interrelations between obligatory (O), permitted ($Pp = \neg O \neg p$) and forbidden ($Fp = O \neg p$) the matrices for the latter two operators can be constructed out of Op by applying the matrix for negation: P turns the basic matrix of p 1 2 3 4 5 6 into the matrix of Pp: 1 3 6 2 1 4. And F respectively into the matrix of Fp: 6 4 1 5 6 3.

Properties of the deontic system 6-DL:

1. The properties 1)-9) of RMQ (section 2. above) hold also for the system 6-DL.

2. The strict version of the system 6-DL, i.e. the strictly valid theorems of 6-DL (those which have a matrix where the highest value $cv = 2$) avoid most of the paradoxes of Deontic Logic (see section 5. below).

3. 6-DL is not suitable for an iteration of the deontic operators or for mixtures (applications of different operators to each other, see section 6. below).

4. The classical (material) version of the system 6-DL, i.e. the classically (materially) valid theorems of 6-DL (those which have a matrix where the highest value $cv = 3$) can be represented by an axiom system (see section 6. below).

5 Theorems of the deontic system 6-DL

Definitions: $Pp = \neg O \neg p$; $Fp = \neg Pp = O \neg p$

5.1

The following theorems hold strictly (are strictly valid) in 6-DL; that is, there $cv = 2$. This is indicated by $\Rightarrow$ or $\Leftrightarrow$.

1. $Op \Rightarrow \neg O \neg p$
2. $Op \Rightarrow Pp$
3. $Pp \Leftrightarrow \neg Fp$
4. $Pp \Leftrightarrow \neg O \neg p$
5. $Fp \Leftrightarrow O \neg p$
6. $O(p \rightarrow q) \Rightarrow (Op \rightarrow Oq)$
7. $O(p \rightarrow q) \Rightarrow (Pp \rightarrow Pq)$
8. $O(p \rightarrow q (\Rightarrow P(p \rightarrow q)$
9. $O(p \vee q) \Rightarrow P(p \vee q)$
10. $O(p \vee q) \Rightarrow Pp \vee Pq$

11. $(Op \lor Oq) \Rightarrow (Pp \lor Pq)$
12. $P(p \lor q) \Rightarrow (Pp \lor Pq)$
13. $(Op \land Oq) \Rightarrow O(p \land q)$
14. $(Op \land Oq) \Rightarrow O(p \lor q)$
15. $(Op \land Oq) \Rightarrow (Op \lor Oq)$
16. $(Op \land Oq) \Rightarrow (Pp \land Pq)$
17. $(Op \land Oq) \Rightarrow P(p \land q)$
18. $(Op \land Oq) \Rightarrow P(p \lor q)$
19. $(Op \land Oq) \Rightarrow (Pp \lor Pq)$
20. $(Op \land \neg Oq) \Rightarrow P(p \land \neg q)$
21. $O((p \land q) \Rightarrow (Op \lor Oq)$
22. $O((p \land q) \Rightarrow P(p \land q)$
23. $O((p \land q) \Rightarrow (Pp \lor Pq)$
24. $(Pp \land Pq) \Rightarrow (Pp \lor Pq)$
25. $(Pp \land Pq) \Rightarrow P(p \lor q)$
26. $(Pp \land \neg Pq) \Rightarrow P(p \land \neg q)$

The proof for these theorems consists of a decision method by checking the cv (highest value) of the matrix of the respective proposition (formula). If the cv = 2, the respective proposition (formula) is strictly valid. This is the case for all the listed theorems 1.-26.

5.2

The following theorems hold only classically or materially (cv = 3), but are strictly invalid in 6-DL.

1. $(Op \lor Oq) \to P(p \lor q)$
2. $(Op \land Oq) \to (Op \to Oq)$
3. $(Op \land Oq) \to P(p \to q)$
4. $O(p \land q) \to P(p \to q)$
5. $O(p \land q) \to (Pp \land Pq)$
6. $(Pp \land Pq) \to (Pp \to Pq)$
7. $P(p \land q) \to (Pp \lor Pq)$
8. $Op \to (Op \lor Oq)$
9. $Pp \to (Pq \lor Pq)$
10. $\neg p \to (p \to Oq)$

11. Principle of Inheritance: If $p \to q$ is valid (or strictly valid), then so is $Op \to Oq$. This principle is invalid in 6-DL.

9. and 10. Are the paradoxes of Free Choice and of Commitment. They are avoided if only the strictly valid part of 6-DL is used as a deontic system (see section 5 below). Similarly for the application to empirical sciences only the strictly valid part of RMQ is used as the applied logical system, since in this case the well-known respective paradoxes can be avoided.

6 Deontic paradoxes avoided

Most of the known deontic paradoxes are invalid in 6-DL. Two of them (Free Choice and Commitment, see above 4.2, 9. and 10.) are strictly invalid, though materially (classically) valid. Thus the strictly valid part of 6-DL avoids all of the following well-known paradoxes of Deontic Logic:

1. Ross Paradox:
   $Op \to O(p \lor q)$ invalid, $cv = 6$

2. Paradox of Derived Obligation
   $O\neg p \to O(p \to q)$ invalid, $cv = 6$

3. Paradox of the Good Samaritan
   $O\neg p \to O\neg(p \land q)$ invalid, $cv = 6$

4. Paradox of Free Choice
   $Pp \to (Pp \lor Pq)$ strictly invalid, $cv = 3$

5. Paradox of Commitment
   $\neg p \to (p \to Oq)$ strictly invalid, $cv = 3$

6. Forrester-Paradox
   We distinguish three versions

   (a) $[Fp \land (p \to Oq) \land p \land (q \to p)] \to (Op \land O\neg p)$ invalid $cv = 5$
   (b) $[Fp \land (p \to Oq) \land p \land O(q \to p)] \to (Op \land O\neg p)$ strictly valid
   (c) $[Fp \land (p \to Oq) \land p \land (Oq \to Op)] \to (Op \land O\neg p)$ strictly valid.

   In the version 6a. the paradox is avoided since the premises are not contradictory such that the contradictory obligation does not follow. Concerning versions 6b. and 6c. however it is easily seen that the premises are contradictory just by applying modus ponens (6c.) or modus ponens and the distribution of O on $\to$ (6b.) which is accepted in almost all systems of Deontic Logic. Thus 6b. and 6c. can hardly be called paradoxes, otherwise every simple example with transparent contradictory premises could be called a paradox.

7. Chisholm-Paradox
   We distinguish four versions

   (a) $[Op \land (p \to Oq) \land \neg p \land (\neg p \to O\neg q)] \to (Oq \land O\neg q)$ invalid, $cv = 5$

(b) $[Op \land O(p \to q) \land \neg p \land O(\neg p \to O\neg q)] \to (Oq \land O\neg q)$ invalid, $cv = 5$

(c) $[Op \land (p \to Oq) \land \neg p \land O(\neg p \to O\neg q)] \to (Oq \land O\neg q)$ invalid, $cv = 5$

(d) $[Op \land O(p \to q) \land \neg p \land (\neg p \to O\neg q)] \to (Oq \land O\neg q)$ strictly valid, $cv = 2$

In all three versions 7a, 7b, 7c the paradox is avoided since in these cases the premises together are contingent and not contradictory. In the case 7d however, the premises are contradictory which can be easily seen by just applying modus ponens and the distribution of O on $\to$ (to get Oq and $O\neg q$). Therefore 7d can hardly be called a paradox because the premises are so apparently and obviously obligation-contradictory.

7 Iterations of deontic operators and axiomatisation

7.1

It can be shown that the deontic system 6-DL is not suitable for iterations of the deontic operators:

$$OOp \to \neg p \text{ strictly valid}, sv = 2$$

Furthermore the deontic system 6-DL is also not suitable for mixed deontic operators:

$$(OPp \land Op) \to p \text{ strictly valid}, cv = 2$$
$$(p \land O\neg p) \to POp \text{ strictly valid}, cv = 2$$
$$(Pp \land \neg p) \to OOp \text{ materially valid}, cv = 3 \text{ (strictly invalid)}$$

This is a drawback only if iterations or mixed operators are really needed. Their interpretation is partially controversial. As Asher and Bonevac say: "The scheme we developed handles arguments without nested deontic operators. It is adequate for almost all the arguments discussed in the literature on deontic logic or practical reasoning."[9] A respective matrix system which can incorporate iterations and mixed operators is possible only by extension of truth-values. For two operations (iteration or mixed) at least 8 values are necessary instead of six. To see that the above difficulties can be avoided by introducing 8 values (0123 for true and 4567 for false), consider the following matrices for obligatory, permitted and forbidden (satisfying the usual interdefinability relations):

Op: 03651463, Pp: 41362140, Fp: 36415637.

The matrices for $\neg, \lor, \land, \to$ are the same and 0 and 7 behave like 1 and 6 when connectives are applied. Then the above four implications and others of that sort which use two operations (iteration or mixed) all become invalid.

However, to introduce an independent action operator seems to be at least as important as to make an extension with iterated or nested operators. But this has to wait for further research.

---

[9] Asher and Bonevac (1997), p. 179. As a matter of fact none of the systems presented in the 14 papers of the book "Defeasible Deontic Logic", ed. by D. Nute, uses iterated or nested deontic operators.

7.2

Albert Anglberger[10] has shown that the classically (materially) valid theorems ($cv = 3$) of 6-DL can be finitely axiomatized. This system (the materially valid theorems of 6-DL) is however not suitable as a deontic logic, since it does not avoid all of the deontic paradoxes (see 9. and 10. of paragraph 4.2 above) although it avoids already 5 of 7 paradoxes. To find axioms for the strictly valid theorems of 6-DL will be a more complicated task for future research.

Acknowledgements

I would like to thank Albert Anglberger and two anonymous referees for several valuable suggestions concerning an earlier version of the paper.

BIBLIOGRAPHY

[1] A. J. J. Anglberger, Dynamic Deontic Logic and its Paradoxes. *Studia Logica* 89: 427-435, 2008.
[2] A. J. J. Anglberger, An Axiomatization of Paul Weingartner's 6-valued Deontic Logic and a Result Concerning its Possible Extensions. Contributed Paper delivered to the International Congress for Logic, Methodology and Philosophy of Science, Nancy, July, 2011.
[3] N. Asher, N. and D. Bonevac, Common Sense Obligation. In: Nute, D. (ed.)(1997), pp. 159-203.
[4] B. F. Chellas, *Modal Logic*. Cambridge: Cambridge Univ. Press, 1980.
[5] H. Czermak, Eine endliche Axiomatisierung von SS1M. In: Morscher, E. et al. (eds.): *Philosophie als Wissenschaft. Essays in Scientific Philosophy*. Comes Verlag, Bad Reichenhall, 1981, p. 245-257.
[6] L. Goble, Murder Most Gentle: The Paradox Deepens. *Philosophical Studies* 64: 217-227, 1991.
[7] P. L. Mott, On Chisholm?s Paradox. *Journal of Philosophical Logic* 2: 197-211, 1973.
[8] D. Nute (ed.), *Defeasible Deontic Logic*. Kluwer, Dordrecht, 1997.
[9] G. Schurz and P. Weingartner, Versimilitude Defined by Relevant Consequence-Elements. A New Reconstruction of Popper?s Original Idea. In: Knipers, Th. (ed.): *What is Closer to the Truth?* . Amstgerdam: Rodopi, p. 47-77, 1987.
[10] P. Suppes, *Probabilistic Metaphysics*. Blackwell, Oxford, 1984.
[11] P. Weingartner, Modal Logics with Two Kinds of Necessity and Possibility. *Notre Dame Journal of Formal Logic* 9: 97-159, 1968.
[12] P. Weingartner, Can there be Reasons for Putting Limitations on Classical Logic? In: Humphries, P. (ed.): *Patrick Suppes, Scientific Philosopher*. Vol. 3. Kluwer, Dordrecht, p. 89-124, 1984.
[13] P. Weingartner, Applications of Logic outside Logic and Mathematics: Do such Applications Force Us to Deviate from Classical Logic? In: Stelzner, W. (ed.): *Zwischen traditioneller und moderner Logik*. Mentis, Paderborn, p. 53-64, 2001.
[14] P. Weingartner, Matrix Based Logic for Application in Physics. *The Review of Symbolic Logic* 2: 132-163, 2009.
[15] P. Weingartner, Basic Logic for Application in Physics and its Intuitionistic Alternative. *Foundations of Physics* 40: 1578-1596, 2010a.
[16] P. Weingartner, An Alternative Propositional Calculus for Application to Empirical Sciences. *Studia Logica* 95: 233-257, 2010b.
[17] P. Weingartner, Matrix Based Logic for Avoiding Paradoxes and its Paraconsistent Alternative. *Manuscrito* 34: 365-388, 2011.

---

[10] Albert Anglberger (2011).

The Psychiatrist's Dilemmas
ANNE FAGOT-LARGEAULT

ABSTRACT. Practising psychiatry means having to face conflicting options every day. Is psychiatric knowledge a branch of biology or a branch of psychology (or both)? Should the psychiatric patient be approached as someone who has a disease and wants to be relieved from a painful condition, or else as someone identified with the disease and who needs to be forced out of a cherished condition (or both)? In order to help the patient, should the psychiatrist prescribe drugs, or sessions of psychotherapy (or both)?

1 Introduction

Psychiatry and philosophy have a turbulent history of encroaching upon each other. I remember Patrick Suppes saying with pride: "I am simplistic". He was proud of it. That was 40 years ago. I want to be simplistic here. Psychiatry has to do with both psyche and body. On topics related to psychiatry, I interacted with Suppes at least twice. In the Fall of 2004, we had a "Mind and brain" seminar at Stanford (Philosophy 389: Dagfinn Føllesdal and Jean-Pierre Changeux also cooperated). In the Fall of 2005 Professor Suppes delivered an invited talk at the College de France in Paris on the "Neuropsychological foundations of philosophy", and we had a small seminar with Prs Drs O. Dulac & C. Chiron, who specialize in the surgical treatment of severe epilepsy in children. I know that Suppes is interested in detecting the impact of psychotherapy on the brain. My modest contribution here will be to formulate schematically philosophical questions emanating from a practice of psychiatry at the hospital, in the emergency room, when I innocently wanted to reconcile the horns of a threefold dilemma.

2 The theoretical dilemma

When a medical student decides to specialize in psychiatry, he/she usually does it on the basis of one of his courses, or of some bright research he's heard of – he almost has not seen any real psychiatric patients yet. He has a theoretical view of the discipline that he is enthusiastic about. There is a common joke that says: being attracted to a theory is like being attracted to a woman. You fall in love, ignore the rest, become a believer, and think your belief is science. Let us take an example.

Student A has just discovered a recent paper on autism (Torres *et al.*, 2012). He believes in molecular psychiatry. He assumes that psychiatric disorders, such as infantile autism, have their cause in the brain cells. The field of his research will be on the genetics of autism and of other pervasive developmental disorders. In his publications you find statistical analyses eventually evidencing links between specific genetic polymorphisms and behavioral phenotypes (such as cognitive deficiencies).

Student B has devoured a book by Irvin Yalom (Yalom, 1980). She believes in psychology-oriented psychiatry. She is interested in the historical antecedents of her patient's troubles. Language is her means of communication with her patients. She listens to them, she tries to empathize with their mental and emotional experience. She publishes case reports.

Neither of them will read the other's articles. Eclectism is rare. As a philosopher of science one would like them to communicate and work together. The patients they want to help have both a brain and an emotional experience. Assuming that their doctors must choose between the two sounds absurd. Yet the theoretical options split the community of psychiatrists. Should we think, as some do, that the unity of the discipline is a mere legacy of the past? The authors of a recent textbook announce that "in order to prevent future psychiatry from dissolving in a number of methodically defined subunits, and to further strenghten person-centered diagnostic approaches, we strongly need the historical perspective" (Paul, 2009, p.12).

3 The relational dilemma

The psychiatry student eventually becomes a resident. How does he/she relate to the mentally or emotionally perturbed patient, for example in the emergency room of a general hospital?

Resident C addresses the patient as a person, who is suffering from an ailment that she is unable to deal with by herself, and the fact that she is here means that she is asking for help. The doctor tries to find out what kind of help is appropriate, he proposes possible solutions, and he lets the patient express preferences about what she thinks is best for her. Should the patient be in severe distress and/or unable to make sound decisions, resident C warns her that he himself will make the decision to institutionalize her now. That is at the risk of seeing the person run away before coercive measures can be taken.

Resident D is convinced that the important point is to make a sharp diagnosis, from which the therapeutic solution necessarily follows. She does not bother asking the patient what he wishes. The patient it out of his mind, he is intrisincally sick, discussing with him is a waste of time. She will have to choose for him. She does not talk to the patient. She talks about the patient to nurses, prescribes whatever drug or brace she deems necessary, and makes a decision: the patient will be locked in a psychiatric wards, or he will be put in a bed until he has sobered up, etc.

Saying that someone has a disease is very different from saying that someone is insane. Let the ailment be, for example, bipolar disorder. The relational dilemma (treating the patient as an equal partner or as someone to be protected and straightened up) refers to a duality of metaphysical presuppositions. Resident C assumes that the patient as a person does not identify with the disease: she has a bout of depression, she would enjoy getting rid of it, in no way can she be deemed responsible for being unwell. Resident D presumes that deep inside the patient is (ontologically) depressed, wants to be depressed, cherishes his depression, coincides with his depressed condition, and that it is nonsense to expect that he might empathize with the doctor's rational project to cure him. Whether the deep consent to his condition may be traceable to his genes, or to an existential choice, the patient here will be deemed to adhere to his depression, as though he were responsible for it; he must be sanctioned for his erroneous choice. This is reminiscent of an old debate in philosophy around the "initial choice of character".[1] Henry Ey qualified mental illness as a "pathology of liberty" (Henri, 2010).

Following Ludwig Binswanger and his notion of Daseinsanalyse, some psychiatrists bet on the possibility for psychoanalysis to have a patient return to the source of his very early traumas and life choice, realize that the melancholic posture he locked himself into was mistaken, opt out of his initial choice, and engage in a new existential attitude.

4 The therapeutic dilemma

Until 1949 there were practically no specific treatments of mental illness, except a few tentative shock treatments: malariatherapy (von Jauregg, 1917), insulin coma (Sakel, in 1934), electroshock (Cerletti & Bini, in 1938); or radical surgery like frontal lobotomy (Moniz, in 1936). Then, within one decade (1949-1959), all potent psychotropic agents were discovered: antipsychotic (or neuroleptic) drugs, antidepressants, benzodiazepine compounds (anxiolytics). From then on, there is virtually no consultation in psychiatry that does not end with a prescription of some chemical. Therefore it might seem that there is no therapeutic dilemma. Prescribing drugs to sedate anxiety, stabilize the mood, or cast out delusions, is standard practice.

The Swiss psychiatrist Eugen Bleuer (contemporary with Sigmund Freud), who identified schizophrenia, and was the director of the Burghölzli clinic in Zürich, did not have any useful drugs to treat his patients. He is known for saying about them: "I couldn't restore them to health – then I listened to them". He meant that he was learning from them about the nature of their ailments. He did not claim that listening to them had any curative power. Freud however believed in the therapeutic efficiency of being able to express unconscious conflicts in language, although he did not disregard the possible effects of chemicals (he used cocaine himself, and studied its effects). Analysing one?s dreams and memories in the course of a psychotherapy has become a routine way for psychiatric patients to be helped. As R.J. Kahana observed half a century ago: "In the past forty years, largely under the impact of psychoanalysis, dynamic psychotherapy has become the principal and essential

---

[1]Plato, *The Republic*, Book X; Aristotle, *Nichomachean Ethics*; Kant, *Critique of the Practical Reason*; A. Adler, *Über den nervösen Character*; J. P. Sartre, *L'Être et le Néant / Being and Nothingness*. For an overview, see entry 'character' in: M. Canto-Sperber *et al.*, Dir., *Dictionnaire d'éthique et de philosophie morale / Diccionario de etica y de filosofía moral / Dictionary of ethics and moral philosophy*, 4th ed rev & augt, Paris, PUF, 1996 / Spanish transl.: Fondo de Cultura Economica USA 2001.

curative skill of the American psychiatrist and, increasingly, a focus of his training" (Kahana, 1998, p. 458).

"The one hand does not know what the other is doing": as a matter of fact, psychiatrists commonly prescribe both psychotherapy and medication to the same patients, with no clear idea of how they interfere. Small empirical evidence has indeed been collected in favor of the belief that both together are better than either one alone. The mechanism is unknown. Note that psychotherapy in general (including cognitive remediation), being not a 'medical' treatment proper, is in many countries not covered by health insurance, while chemicals are. That makes it an elitist choice. The psychoanalytic school has argued that the high cost of psychoanalysis is a decisive element in its effectiveness. How such an hypothesis may coexist with biological hypotheses on the causation of mental illness remains a mystery.

The psychiatrist Edouard Zarifian deplored the 'religious war' that was going on in the nineteen eighties between exclusive believers in psychoanalysis ("brainless psychiatry of the past") and fanatic supporters of neurobiology ("mindless psychiatry of the future") (Zarifian, 1988, p.204).

The war has now calmed down, the dissociation remains, as if psychiatrists could only jump from one side to the other. Zarifian himself went from biological psychiatry, of which he was a specialist, to social and psychological psychiatry, when he was made responsible of the mental health of a large district. The psychiatrist Eric Kandel, Nobel prize in medicine 2000, went in the other direction. Here is the way how, in his autobiography, he summarizes his successive choices: "I entered Harvard to become a historian and left to become a psychoanalyst, only to abandon both of those careers to follow my intuition that the road to a real understanding of mind must pass through the cellular pathways of the brain" (Kandel, 2006, p. 429).

BIBLIOGRAPHY

[1] R. J. Kahana, Psychotherapy: models of the essential skill. In E. R. Kandel, 'A new intellectual framework for psychiatry', *American Journal of Psychiatry*, 155: 457-469, 1998.
[2] E. R. Kandel, *In Search of Memory. The Emergence of a New Science of Mind*. New York: Norton, 2006.
[3] E. Henri, *Manuel de psychiatrie* (coll. Bernard et Brisset). Paris: Masson, 1960; 7th edition, Elsevier Masson, 2010.(eds.), *Psychiatric Diagnosis. Challenges and prospects*. Oxford: Wiley-Blackwell, 2009.
[4] A. R. Torres, J. B. Westover and A. J. Rosenspire, HLA Immune Function Genes in Autism, *Autism Research and Treatment*, 2012. Published online 2012 February 15. doi: 10.1155/2012/959073.
[5] E. Zarifian, *Les jardiniers de la folie*. Paris: Odile Jacob, 1988.
[6] I. D., Yalom, *Existential Psychotherapy*. New York: Basic Books, 1980.

www.ingramcontent.com/pod-product-compliance
Lightning Source LLC
Chambersburg PA
CBHW070733160426
43192CB00009B/1423